BASIC CONSTRUCTION MATH REVIEW

A Manual of Basic Construction Mathematics for Contractor and Tradesman License Examinations

Atlas Publishing

specializes in providing Bookstores, Trade/Vocational Schools, Colleges and Universities, Retail Chains, Specialty Retailers and Independent Hardware Stores with the educational resources needed to promote continued learning. We carry a complete line of code books and references as well as training DVD's and Videos, for the electrical, plumbing, building and construction trades.

ATLAS PUBLISHING

30 Oser Avenue, Suite 500
Hauppauge, NY 11788

Phone 888.226.7052
Fax 800.719.4402
www.atlaspublishing.com

Preface

Construction competancy tests are usually book tests. An understanding of formulas, equations, areas and volumes and other math fundamentals is essential if many of the problems on these tests are to be solved correctly.

This manual was written to help you learn or re-learn these arithmetic, algebra and geometry fundamentals. Some mathematicians might say that all of the "usual" basic math topics are not covered in the following pages — and they're right. Only those math basics that will help you pass that competancy test are discussed in this manual.

Self-study with any book requires much discipline and diligence. Some time-proven tips have been listed on page *vii* to help you know how to study and gain the most benefit for your time invested.

With this manual and a reasonable effort, you will soon have that good understanding of basic math.

Good luck!

ATLAS PUBLISHING

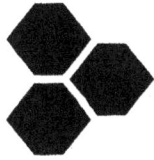

Table of Contents

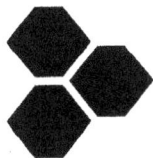

How to Study and Test Strategy

Before the Test

- Develop a positive attitude about the exam — a negative attitude will only hinder your learning.

- Study with friends occasionally, exchange ideas, cover the topics most unfamiliar to you.

- Study regularly. You will learn best if you take it a little at a time. Repetition will help to allow the concepts to become second nature to you both at the exam and in the field.

- Before beginning to study, use a moment to relax and clear your mind from all the events of the day.

At the Test

- Listen closely to the test administrator. If you don't understand the directions you have heard, ask the proctor.

- Read the directions carefully. Don't assume that you know what the exam writer is going to say in the instructions.

- Briefly glance through the entire test. Get an idea of how much time per question is allowed.

- Understand the scoring method to be used. If guessing is advisable, by all means, guess at those answers that you are not sure of.

- Budget your time. You should not have worked half the problems when half of the allotted time is gone.

- Do the easy questions first. You'll gain confidence, you'll get credit for answers you're most sure of, and you'll give your subconscious a chance to work on the tough ones. Sometimes answers to problems will appear in subsequent related questions.

- Think!

- Look over your test and answers. You may find some mistakes that can be corrected easily. Proofread to ensure that you have answered all questions and that the answers are in the right place.

- It should be noted, on most tests your first answer has the highest probability to be correct. However, math exams are obviously different from "information" based exams and checking your work is an excellent way to improve your score.

1 Fractions

A <u>fraction</u> is a part of any whole number, such as $^1/_2$, $^5/_{16}$ or $^9/_{32}$.

The <u>numerator</u> is the part of a fraction written above the line, such as the 3 in $^3/_4$, $^3/_8$ and $^3/_{32}$.

The <u>denominator</u> is the part of a fraction written below the line, such as the 7 in $^1/_7$, $^3/_7$ and $^9/_7$.

A <u>proper fraction</u> has a value less than 1 (one). The numerator is a smaller number than the denominator, such as $^4/_9$ and $^{15}/_{16}$.

An <u>improper fraction</u> has a value greater than 1 (one). The numerator is a larger number than the denominator, such as $^8/_3$, $^3/_2$ and $^{21}/_{17}$.

A <u>mixed number</u> consists of a whole number and a fraction written together, such as 1 $^5/_8$ or 23 $^9/_{16}$.

A <u>whole number</u> is a digit from 0 to 9 or a combination of digits, such as 15, 528, etc.

> *When the numerator and denominator of a fraction are both multiplied or divided by the same number, the value of the fraction remains unchanged.*

EXAMPLES:

$$\frac{1}{2} = \frac{1 \times 3}{2 \times 3} = \frac{3}{6} = \frac{3 \times 10}{6 \times 10} = \frac{30}{60} = \frac{30 \div 30}{60 \div 30} = \frac{1}{2}$$

$$\frac{40}{100} = \frac{40 \div 10}{100 \div 10} = \frac{4 \div 2}{10 \div 2} = \frac{2}{5} = \frac{2 \times 20}{5 \times 20} = \frac{40}{100}$$

REDUCTION OF FRACTIONS

When a fraction has the smallest possible whole numbers for its numerator and denominator, the fraction is said to be in its "lowest form" or "lowest terms".

AT A GLANCE

$\dfrac{4}{4}$ ◄ **Numerator**
  ◄ **Denominator**

$\dfrac{35}{17}$
↑
Improper Fraction

$2\tfrac{3}{4}$
↑
Mixed Number

18
↑
Whole Number

$\dfrac{15}{7} = 2\tfrac{1}{7}$
↑
Lowest Terms

The fraction $^6/_{10}$ is not in its lowest form, since 2 will divide evenly into both 6 and 10 ([6 ÷ 2]/[10 ÷ 2] = $^3/_5$). Thus, 3/5 is in its lowest form, since no other whole number will divide evenly into both the numerator and denominator.

> *To reduce a fraction to its lowest terms, divide both the numerator and denominator by the largest whole number that will divide both exactly.*

EXAMPLE:

Reduce $^{10}/_{24}$ to its lowest form.

SOLUTION:

$$\frac{10}{24} = \frac{10 \div 2}{24 \div 2} = \frac{5}{12}$$

If you do not immediately see the largest whole number that will divide into both numerator and denominator, reduce the fraction by repeated steps. If both numerator and denominator are even, start by halving both until you develop the skills in spotting the greatest common factor.

EXAMPLE:

Reduce $^{128}/_{288}$ to its lowest form.

SOLUTION:

$$\frac{128 \div 4}{288 \div 4} = \frac{32 \div 2}{72 \div 2} = \frac{16 \div 4}{36 \div 4} = \frac{4}{9}$$

A fraction that has the same numerator as denominator reduces to one (1).

EXAMPLE:

$$\frac{25}{25} = \frac{25 \div 5}{25 \div 5} = \frac{5}{5} = \frac{5 \div 5}{5 \div 5} = \frac{1}{1}$$

PRACTICE SET #1

Reduce to lowest terms

1. $^6/_8 =$

2. $^{10}/_{12} =$

3. $^{15}/_{25} =$

4. $^7/_{98} =$

5. $^{144}/_{264} =$

6. $^9/_{54} =$

7. $^{45}/_{450} =$

8. $^4/_{1000} =$

9. $^{42}/_{126} =$

10. $^{99}/_{1089} =$

11. $^{482}/_{482} =$

12. $^9/_{122} =$

CHANGING IMPROPER FRACTIONS AND MIXED NUMBERS

To change an improper fraction to a whole or mixed number, divide the numerator by the denominator and put the remainder (if any) over the denominator.

EXAMPLE: Reduce $^{14}/_3$ to its lowest form.

STEP 1	**STEP 2**	**STEP 3**	**STEP 4**
$\dfrac{14}{3} = 3\overline{\smash{)}14}$	$\begin{array}{r} 4 \\ 3\overline{\smash{)}14} \\ -12 \\ \hline 2 \end{array}$ 3 goes into 14 4 times...	2 is the remainder so...	$\dfrac{14}{3} = 4\frac{2}{3}$

EXAMPLE: Change $^{635}/_8$ to a mixed number.

STEP 1	**STEP 2**	**STEP 3**	**STEP 4**
$\dfrac{635}{8} = 8\overline{\smash{)}635}$	$\begin{array}{r} 79 \\ 8\overline{\smash{)}635} \\ 56 \\ \hline 75 \\ 72 \\ \hline 3 \end{array}$ 8 goes into 635 79 times...	3 is the remainder so...	$\dfrac{635}{8} = 79\frac{3}{8}$

To change a mixed number to an improper fraction, multiply the denominator by the whole number, add the numerator to this product, and place the sum over the denominator.

EXAMPLE: Reduce $4\,^7/_8$ to an improper fraction.

SOLUTION:

$$4\,^7/_8 = \frac{8 \times 4 + 7}{8} = \frac{32 + 7}{8} = \frac{39}{8}$$

EXAMPLE: Reduce $11\,^3/_5$ to an improper fraction.

SOLUTION:

$$11\,^3/_5 = \frac{(5 \times 11) + 3}{5} = \frac{55 + 3}{5} = \frac{58}{5}$$

ADDITION OF FRACTIONS AND MIXED NUMBERS

To add fractions that have the same denominators,
add the numerators and place over the denominator.

EXAMPLE: Add $2/15$, $4/15$ and $7/15$.

SOLUTION:

$$\frac{2 + 4 + 7}{15} = \frac{13}{15}$$

Just as you cannot add inches and feet without first changing both to a common unit of measure, you cannot add or subtract unlike fractions until you have first changed them to a lowest common denominator (unlike fractions are fractions with different denominators).

This lowest common denominator (let's abbreviate it as LCD) is the smallest number that all the denominators will divide into exactly. 6 is the LCD for $1/2$ and $1/3$, 12 is the LCD of $1/3$ and $1/4$, and 45 is the LCD of $1/9$ and $1/5$.

$$\frac{1}{②}, \frac{1}{③}$$

LCD = 6

$$\frac{1}{3}, \frac{1}{4}$$

LCD = 12

$$\frac{1}{9}, \frac{1}{5}$$

LCD = 45

To add unlike fractions, first change them to equivalent fractions (with the same denominator), then add the numerators and place the sum over the common denominator.

EXAMPLE: Add $^1/_2$ and $^1/_3$.

SOLUTION:

The smallest number that 2 and 3 can be divided into evenly is 6. The number 6 is the LCD and each fraction to be added must be written with 6 as the denominator.

$$\frac{1}{2} = \frac{3}{6} \text{ and } \frac{1}{3} = \frac{2}{6}, \text{ so}$$

$$\frac{3}{6} + \frac{2}{6} = \frac{3+2}{6} = \frac{5}{6}$$

EXAMPLE: Add $^1/_4$ and $^5/_6$.

SOLUTION:

The LCD is 12, so rewriting each fraction with 12 as the denominator:

$$\frac{1}{4} = \frac{3}{12} \text{ and } \frac{5}{6} = \frac{10}{12}, \text{ so}$$

$$\frac{3}{12} + \frac{10}{12} = \frac{3+10}{12} = \frac{13}{12} \quad \text{We then reduce this to its simplest form and it becomes } 1\,^1/_{12}$$

EXAMPLE: Add $^1/_3$, $^1/_4$ and $^1/_5$.

SOLUTION:

The LCD is not so obvious in this problem. Try doubling the largest denominator. $5 \times 5 = 25$ but 3 and 4 will not evenly divide into 25. Try multiplying two of the denominators $3 \times 4 = 12$, no good. $3 \times 5 = 15$, no good. $4 \times 5 = 20$, no good. As a last resort, multiply all of the denominators. $3 \times 4 \times 5 = 60$ and 60 is the LCD. So...

$$\frac{1}{3} + \frac{1}{4} + \frac{1}{5} = \frac{20}{60} + \frac{15}{60} + \frac{12}{60} = \frac{20+15+12}{60} = \frac{47}{60} \quad \text{OR}$$

$$\frac{1}{3} = \frac{1 \times 20}{3 \times 20} = \frac{20}{60}$$

$$\frac{1}{4} = \frac{1 \times 15}{4 \times 15} = \frac{15}{60} \quad\longrightarrow\quad \frac{20}{60} + \frac{15}{60} + \frac{12}{60} = \frac{47}{60}$$

$$\frac{1}{5} = \frac{1 \times 12}{5 \times 12} = \frac{12}{60}$$

If you are having trouble writing $^1/_3$ with 60 as a denominator in the problem above, for example, just divide the smaller denominator into the larger and multiply that result by the numerator.

EXAMPLE: Rewrite $^2/_3$ with 12 as a denominator.

SOLUTION:

$$3\overline{)12}^{\ 4}\ , \text{ so } 4 \times 2 = 8$$
$$\underline{12}$$

$$\frac{2}{3} = \frac{2 \times 4}{3 \times 4} = \frac{8}{12}$$

EXAMPLE: Rewrite $^5/_7$ with 56 as a denominator.

SOLUTION:

$$7\overline{)56}^{\ 8}\ , \text{ so } 8 \times 5 = 40$$
$$\underline{56}$$

$$\frac{5}{7} = \frac{5 \times 8}{7 \times 8} = \frac{40}{56}$$

Finding the lowest common denominator (LCD) will come easier with practice.
Let's work one more problem.

EXAMPLE: Add $^2/_5$, $^1/_{10}$ and $^4/_9$.

SOLUTION:

The LCD is 90. Rewriting,

$$\frac{2}{5} = \frac{2 \times 18}{5 \times 18} = \frac{36}{90}$$

$$\frac{1}{10} = \frac{1 \times 9}{10 \times 9} = \frac{9}{90}$$

$$\frac{4}{9} = \frac{4 \times 10}{9 \times 10} = \frac{40}{90}$$

$$\frac{36}{90} + \frac{9}{90} + \frac{40}{90} = \frac{85}{90}$$

To summarize the steps to finding the LCD,

1) Examine the fractions to see if any existing denominator will work as LCD.
2) If none of the denominators will serve as the LCD, try doubling or tripling the largest denominator.
3) If no LCD is found, then multiply the denominators by each other. The resulting common denominator may have to be reduced to be the LCD. This will not necessarily result in the smallest LCD, but will ALWAYS work.

To add mixed numbers, add the fractions separately and add the result to the sum of the whole numbers.

EXAMPLE: Add $2\,^3/_8$ and $1\,^3/_4$.

SOLUTION:

$$2\,^3/_8$$
$$+\ 1\,^6/_8$$
$$\overline{3\,^9/_8} = 3 + 1\,^1/_8 = 4\,^1/_8$$

PRACTICE SET #3

Add the following fractional quantities then reduce the answer to simplest form:

1. $^3/_4 + ^3/_4 =$

2. $^5/_8 + ^7/_8 =$

3. $^1/_2 + ^1/_4 =$

4. $^1/_3 + ^1/_9 =$

5. $^3/_{10} + ^1/_{20} =$

6. $^3/_{10} + ^1/_{20} + ^7/_{30}$

7. $^5/_{13} + ^7/_{26} + ^{11}/_{52} =$

8. $5\,^1/_2 + 2\,^1/_2 =$

9. $7\,^1/_4 + 3\,^2/_3 =$

10. $15\,^1/_9 + 8\,^1/_3 + 2\,^1/_{27} =$

11. A piece of lumber $3\,^1/_2$" thick is covered on both sides by $^3/_4$" plywood. What is the total thickness?

12. A carpenter used random widths of knotty pine to panel a room. Starting in one corner, he used pieces that were $4\,^5/_8$, $7\,^5/_8$, $9\,^3/_8$ & $5\,^5/_8$ inches wide. How wide was the wall panel at that time?

Continued on next page

13. What space on a bolt will five washers of the following thicknesses occupy: $^1/_{16}$, $^3/_{32}$, $^3/_{16}$, $^3/_{64}$ and $^5/_8$ inches?

14. Six resistances are connected in a single series circuit. If the resistance in a series circuit is the sum of all resistances in that circuit, what is the total amount of resistance in the circuit with resistances of $^2/_3$, $^1/_3$, $^5/_6$, $^1/_4$, $^2/_5$ and $^3/_5$ ohms?

15. What is the total horsepower of following motors: 1/16, 3/4, 1/2, 1/6 and 1/3 h.p.?

16. A lead pipe installation is made up of 13 $^1/_2$ lbs. of 2" pipe, 18 $^2/_3$ lbs. of 1 $^1/_2$" pipe, and 11 $^1/_8$ lbs. of 1" pipe. What is the total weight of lead pipe used?

SUBTRACTION OF FRACTIONS AND MIXED NUMBERS

To subtract like fractions, subtract the numerators and place the difference over the common denominator.

EXAMPLE: Subtract $^3/_8$ from $^5/_8$.

SOLUTION:

$\frac{5}{8} - \frac{3}{8} = \frac{5-3}{8} = \frac{2}{8}$ which simplifies to $\frac{1}{4}$

EXAMPLE: Subtract $^3/_{16}$ from $^{15}/_{16}$.

SOLUTION:

$\frac{15}{16} - \frac{3}{16} = \frac{15-3}{16} = \frac{12}{16}$ which simplifies to $\frac{3}{4}$

To subtract unlike fractions, find the lowest common denominator, rewrite and proceed as for subtraction of like fractions.

EXAMPLE: Subtract $^1/_8$ from $^3/_4$.

SOLUTION:

$\frac{3}{4} = \frac{6}{8}$, so $\frac{6}{8} - \frac{1}{8} = \frac{6-1}{8} = \frac{5}{8}$

EXAMPLE: Subtract $^2/_9$ from $^1/_4$.

SOLUTION:

36 is the LCD, so $^1/_4 = {}^9/_{36}$ and $^2/_9 = {}^8/_{36}$

$$\frac{9}{36} - \frac{8}{36} = \frac{9-8}{36} = \frac{1}{36}$$

To subtract mixed numbers, first find the difference between the fractions, then find the difference between the whole numbers. If the denominators are different, it will be necessary to find the LCD and rewrite the fractions.

EXAMPLE: Subtract $2\ ^3/_{16}$ from $4\ ^9/_{16}$.

SOLUTION:

$$
\begin{array}{r}
4\ ^9/_{16} \\
-\ 2\ ^3/_{16} \\
\hline
2\ ^6/_{16} = 2\ ^3/_8
\end{array}
$$

EXAMPLE: Subtract $6\ ^7/_8$ from $15\ ^1/_8$.

SOLUTION:

The denominators are the same, but you cannot take $^7/_8$ from $^1/_8$. "Borrow" one unit ($^8/_8$) from the whole number 15 and add it to the $^1/_8$.

$$
\begin{array}{rcl}
15\ ^1/_8 & = & 14\ ^9/_8 \\
-\ 6\ ^7/_8 & = & -\ 6\ ^7/_8 \\
\hline
& & 8\ ^2/_8 = 8\ ^1/_4
\end{array}
$$

PRACTICE SET #4

Subtract the following quantities:

1. $^1/_2 - {}^1/_2 =$

2. $^3/_2 - {}^1/_2 =$

3. $^3/_8 - {}^1/_8 =$

4. $^7/_{16} - {}^1/_8 =$

5. $^7/_8 - {}^1/_2 =$

Continued on next page

6. $^4/_5 - ^3/_{10} =$

7. $^{15}/_{16} - ^3/_4 - ^1/_8 =$

8. $5\ ^5/_6 - 2\ ^1/_6 =$

9. $2\ ^1/_{32} - 1\ ^1/_{64} =$

10. $12\ ^9/_{17} - 5\ ^2/_{34} =$

11. $9\ ^1/_3 - 7\ ^2/_3 =$

12. $16\ ^3/_8 - 9\ ^5/_6 =$

13. $450 - 52\ ^7/_8 =$

14. A loaded truck weighed 8471 $^5/_4$ lbs. The empty truck weighed 3549 $^1/_3$ lbs. What was the weight of the load?

15. How much must a $^7/_8$" thick board be planed if the finished thickness is to be $^{25}/_{32}$" ?

16. By how much does the length of an 8" by 3 $^3/_4$" brick exceed its width?

17. A motor brush is 2 $^1/_8$ inches long. How long is it after $^{49}/_{64}$ inch wears away?

18. The length of threading on a box connector is $^7/_{16}$". The locknut is $^1/_8$" thick and the metal into which this connector is inserted is $^1/_{32}$". How much threading is left for a threaded bushing?

19. From a piece of pipe 34 $^3/_8$" long, three pieces of 6 $^3/_4$", 7 $^3/_{16}$" and 11 $^5/_8$" are cut. How long is the length of pipe that is left?

20. Two fitters work on the same job. Fitter A works 12 $^3/_4$ hours and Fitter B works 13 $^1/_3$ hours. Fitter C does the whole job in 25 $^1/_4$ hours. Does C do the work in less time than A and B? What is the time difference?

MULTIPLICATION OF FRACTIONS AND MIXED NUMBERS

To multiply fractions, multiply the numerators to get the numerator answer, multiply the denominators to get the denominator answer, then reduce to lowest terms.

EXAMPLE:

Multiply $^1/_4$ by $^2/_3$.

SOLUTION:

$$\frac{1 \times 2}{4 \times 3} = \frac{1 \times 2}{4 \times 3} = \frac{2}{12} = \frac{1}{6}$$

EXAMPLE: Find the product of $^3/_8$ and $^7/_8$.

SOLUTION:

$$\frac{3 \times 7}{8 \times 8} = \frac{3 \times 7}{8 \times 8} = \frac{21}{64}, \text{ will not reduce}$$

To multiply a whole number by a fraction, write the whole number as a fraction and multiply as fractions (note: the whole number, say, 4 is the same as 4/1).

EXAMPLE: Multiply 4 by $^3/_{16}$.

SOLUTION:

$$4 = \frac{4}{1} \text{ and } \frac{4}{1} \times \frac{3}{16} = \frac{4 \times 3}{1 \times 16} = \frac{12}{16} = \frac{3}{4}$$

To multiply a fraction by a mixed number, change the mixed number to an improper fraction (numerator larger than denominator) and multiply as for two fractions: each numerator times numerator and the denominator times denominator.

EXAMPLE: Multiply 2 $^1/_4$ by $^1/_3$.

SOLUTION:

$$2 \tfrac{1}{4} = \frac{9}{4} \text{ so,}$$

$$\frac{9}{4} \times \frac{1}{3} = \frac{9}{12} = \frac{3}{4}$$

EXAMPLE: Multiply 6 $^3/_8$ by $^1/_5$.

SOLUTION:

$$6 \tfrac{3}{8} = \frac{51}{8} \text{ so,}$$

$$\frac{51}{8} \times \frac{1}{5} = \frac{51}{40} = 1 \tfrac{11}{40}$$

To multiply two mixed numbers, change both to improper fractions and multiply.

EXAMPLE:

Multiply 5 $^1/_4$ by 7 $^3/_8$.

SOLUTION:

$5 \frac{1}{4} = \frac{21}{4}$, and $7 \frac{3}{8} = \frac{59}{8}$, so

$$\frac{21}{4} \times \frac{59}{8} = \frac{1239}{32} = 38 \ 23/32$$

CANCELLATION

Cancellation is a time-saving way of working out the multiplication of fractions.

Knowing that you can divide the numerator and denominator of a fraction without changing its value, you can take common factors out of the fractions before working the problem.

Compare:

Long way $\dfrac{9}{16} \times \dfrac{4}{3} \times \dfrac{2}{3} = \dfrac{9 \times 4 \times 2}{16 \times 3 \times 3} = \dfrac{72}{144} = \dfrac{6}{12} = \dfrac{1}{2}$

Cancellation

$$\dfrac{\overset{3}{\cancel{9}}}{16} \times \dfrac{4}{\underset{1}{\cancel{3}}} \times \dfrac{2}{3} = \dfrac{\overset{1}{\cancel{3}}}{16} \times \dfrac{4}{1} \times \dfrac{2}{\underset{1}{\cancel{3}}} = \dfrac{1}{\underset{4}{\cancel{16}}} \times \dfrac{\cancel{4}}{1} \times \dfrac{2}{1} = \dfrac{1}{\cancel{4}} \times \dfrac{1}{1} \times \dfrac{\overset{1}{\cancel{2}}}{\underset{2}{}} = \dfrac{1}{2}$$

Remember, cancellation can only be applied to multiplication and division, <u>never</u> to addition and subtraction of fractions.

EXAMPLE:

Multiply $^5/_8$ x $^2/_5$ x $^6/_{12}$ x $^1/_3$.

SOLUTION:

$$\dfrac{\cancel{5}}{\underset{4}{\cancel{8}}} \times \dfrac{\overset{1}{\cancel{2}}}{\cancel{5}} \times \dfrac{\overset{1}{\cancel{6}}}{\underset{6}{\cancel{12}}} \times \dfrac{1}{\underset{1}{\cancel{3}}} = \dfrac{1}{24}$$

PRACTICE SET #5

Multiply the following quantities:

1. $3/5 \times 1/5 =$

2. $4/7 \times 5/8 =$

3. $1\ 2/3 \times 1/3 =$

4. $3/4 \times 2\ 7/8 =$

5. $2/15 \times 5 \times 1/3 =$

6. $7\ 6/11 \times 3\ 2/3 =$

7. $12 \times 637\ 1/2 =$

8. $23\ 7/12 \times 1\ 1/2 =$

9. $2/3 \times 3/4 \times 4/5 \times 7/8 =$

10. $3\ 1/2 \times 5\ 2/3 \times 4\ 4/5 =$

11. There are 15 risers in a stairs. If each riser is $7\ ^{11}/_{16}$" high, what is the total stair height?

12. Thirteen strips, $3/4$" wide each, are to be ripped from a piece of plywood. If $1/8$" is lost with each cut, how much of a plywood sheet is used to make the strips?

13. Shingles are laid so that $5\ 1/8$" are exposed in each layer. How many feet of roof are covered in 31 courses (answer in feet and inches)?

14. A wiring job calls for 30 pieces of conduit $7\ 1/2$ feet long, 6 pieces $5\ 3/4$ feet long, and 11 pieces $2\ 1/2$ feet long. What is the total length of conduit for this job?

15. A motor rotates at the rate of $1757\ 2/3$ revolutions per minute. How many revolutions does the motor make in $2\ 1/2$ minutes?

16. The construction of a building requires 14,700 cubic feet of brick work. The sizes of the brick and joint require $19\ 1/4$ bricks per cubic foot. How many bricks are needed for the job?

17. $1\ 1/2$ tons of pipe fittings were delivered to a job. If $1\ 1/5$ of a ton was used during construction, how many pounds of fittings were left unused (answer in lbs.)?

18. A cubic foot of water contains $7\ 1/2$ gallons. If a gallon of water weighs $8\ 1/3$ lbs., how much does a cubic foot of water weigh?

19. A box contains 556 bolts. If each bolt weighs $2\ ^{7}/_{16}$ pounds and the box weighs $7\ 1/3$ pounds, what is the total weight?

DIVISION OF FRACTIONS AND MIXED NUMBERS

To divide by a fraction, invert the divisor (turn the divisor upside down) and proceed as in multiplication of fractions.

EXAMPLE:

Divide $^3/_8$ by $^1/_4$.

SOLUTION:

$$\frac{3}{8} \div \frac{1}{4} = \frac{3}{8} \times \frac{4}{1} = \frac{12}{8} = 1\,^4/_8 = 1\,^1/_2$$

EXAMPLE:

Divide $^2/_5$ by $^9/_{16}$.

SOLUTION:

$$\frac{2}{5} \div \frac{9}{16} = \frac{2}{5} \times \frac{16}{9} = \frac{32}{45}$$

◇◇ **AT A GLANCE**

The DIVISOR is the number of pieces you will be "cutting" the dividend.

$$4 \div 2$$

Dividend   **Divisor**

$$\frac{1}{2} \div \frac{1}{16}$$

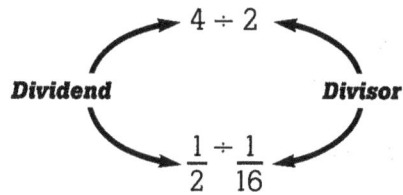

Mixed numbers should be changed to improper fractions before inverting the divisor and multiplying.

EXAMPLE:

Divide $3\,^1/_3$ by $^1/_2$.

SOLUTION:

$3\,^1/_3 = {}^{10}/_3$, so

$$\frac{10}{3} \div \frac{1}{2} = \frac{10}{3} \times \frac{2}{1} = \frac{20}{3} = 6\,^2/_3$$

If the answer to the problem above does not "sound" right, remember, to divide $3\,^1/_3$ by $^1/_2$ means the same as "how many $^1/_2$'s are there in $3\,^1/_3$". As in multiplying fractions on our earlier pages, try to cancel or simplify <u>before</u> you multiply to keep the equations more manageable.

Be careful of the wording of questions. The Divisor is the "cutting" number and the dividend is the piece being "chopped up".

EXAMPLE: Divide $1/8$ into $10 \ 1/4$.

SOLUTION:

$10 \ 1/4 = {}^{41}/_4$, so

$$\frac{41}{4} \div \frac{1}{8} = \frac{41}{4} \times \frac{8}{1} = \frac{328}{4} = 82$$

If the divisor is a whole number, write it over one, invert and multiply.

EXAMPLE: Divide $4/5$ by 7.

SOLUTION:

$$\frac{4}{5} \div 7 = \frac{4}{5} \div \frac{7}{1} = \frac{4}{5} \times \frac{1}{7} = \frac{4}{35}$$ In this example, diving by 7 is the same as multiplying by $1/7$.

PRACTICE SET #6

Divide the following quantities:

1. $3/4 \div 1/2 =$

2. $7/16 \div 7/4 =$

3. $\dfrac{18}{3 \ 1/3} =$

4. $4/7 \div 4/7 =$

5. $2 \ 2/3 \div 1/3 =$

6. $\dfrac{9 \ 5/6}{3} =$

7. $27 \ {}^{11}/_{12} \div 3 \ 1/2 =$

8. $92 \div 13 \ {}^{11}/_{12} =$

9. $276 \ 1/4 \div 7 =$

10. $\dfrac{8 \ 1/2}{2 \ 1/4} =$

Continued on next page

11. How many sheets of metal $1/32$" thick are in a pile 25 $1/2$" high?

12. How many pieces of wire each $2/3$ foot long can be cut from a roll 15 feet in length? How much wire is left after the last piece is cut?

13. A wall rises 5 $1/4$" with each course of brick. How many courses are needed for a wall 105" high?

14. How many 7 $1/2$" risers are there in a stairs 90" high?

15. If 12 $6/7$ watts are distributed equally over 7 resistors, what is the average number of watts per resistor?

16. How many whole wedges 3 $7/8$" long can be made from 30 wedge strips each 36" long?

17. A story pole is to be divided into 39 spaces. If the pole is 107 $1/4$" long, how long is each space?

18. If a water tank holds 684 $1/2$ gallons, what is its cubic foot capacity (7 $1/2$ gallons per cubic foot)?

19. A cubic foot of water weighs 62 $1/2$ pounds and contains 7 $1/2$ gallons. How much does one gallon of water weigh?

20. A barrel of fittings weighs 523 $1/2$ lbs. The barrel weighs 18 $1/4$ lbs. and each fitting weighs 11 $3/4$ lbs. How many fittings does the barrel hold?

◆ 2 *Decimals*

A <u>decimal</u> is a fraction with a denominator of 10 or some multiple of 10. However, the denominators are not written but are indicated by the position of a dot called a decimal point.

Decimals give us a way of writing proper fractions that have denominators ending with zero (0). For example, to write the fraction $^7/_{10}$ as a decimal, we write .7, since the first place to the right of the decimal point is the tenths place.

The following illustration may help you to understand how the position of the decimal point indicates the value of the decimal:

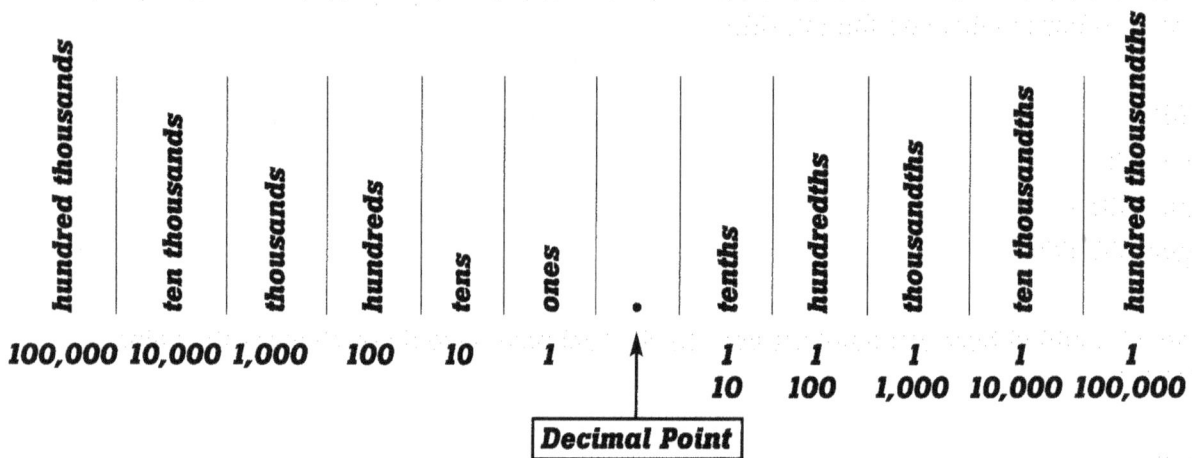

hundred thousands	ten thousands	thousands	hundreds	tens	ones	.	tenths	hundredths	thousandths	ten thousandths	hundred thousandths
100,000	10,000	1,000	100	10	1		$\frac{1}{10}$	$\frac{1}{100}$	$\frac{1}{1,000}$	$\frac{1}{10,000}$	$\frac{1}{100,000}$

Decimal Point

Thus, the fraction 7/10 is written as .7 and is read as "seven tenths".

The fraction 7/100 is written as .07 and is read as "seven hundredths" and so on.

If numbers are on both sides of the decimal point, the number is called a <u>decimal mixed number.</u> The decimal mixed number 3.14 is read "three point one four" or "three and fourteen hundredths". Easy as pi.

Adding zeros to decimals may help reduce errors when adding large columns of numbers. An important fact to remember is... the value of a decimal fraction is not changed if zeros are written at the right end of it, the same as adding zeros in front of a whole number.

EXAMPLES

.7 equals .70

.7 equals .700

.7 equals .700000

Zeros can be added to the left side of the decimal point without changing the value of the decimal if there are no other numbers on the left side.

EXAMPLES

.7 equals 0.7

.7 equals 00.7

.7 equals 00000.7

Zeros cannot be added between numbers and the decimal point — to do so changes the value of the decimal.

EXAMPLES

.3 does not equal .03

3.3 does not equal 300.3

876.23 equals 876.2300 but not 876.023

DECIMAL FRACTION CONVERSION

To change a fraction to a decimal, divide the denominator into the numerator.

EXAMPLE:

Change $\frac{1}{5}$ to a decimal.

SOLUTION:

$$5\overline{)\begin{array}{c} .2 \\ 1.0 \\ \hline 1\,0 \end{array}} = .2$$

AT A GLANCE

It is crucial to keep track of your decimal as you divide. It is placed directly above the dividend in standard division...

$$5\overline{)\begin{array}{c} .2 \\ 1.0 \end{array}}$$

EXAMPLE: Change 2 ¹/₄ to a decimal.

SOLUTION:

2 ¹/₄ = ⁹/₄

$$
\begin{array}{r}
2.25 \\
4\overline{)9.00} \\
\underline{8} \\
10 \\
\underline{8} \\
20 \\
20 \\
\end{array} = 2.25
$$

To change a decimal to a fraction, count the number of places the decimal has to move to the right to make the number "whole". Divide by "1" and the same number of zeros as places you moved the decimal.

EXAMPLE: Change .25 to a fraction.

SOLUTION:

.25   . 2 5 $\frac{25}{100}$ $\overset{Simplify}{=}$ $\frac{5}{20}$ $\overset{Simplify}{=}$ $\frac{1}{4}$

one two   two places, two zeros

EXAMPLE: Change 1.95 to a fraction.

SOLUTION:

$1.95 = \dfrac{195}{100} = \dfrac{39}{20} = 1\ ^{19}/_{20}$

PRACTICE SET #7

Change the following fractions to decimal equivalents:

1. ³/₁₀ =

2. ³/₅ =

3. ³/₈ =

4. ⁷/₁₆ =

5. ⁵/₆₄ =

6. 2 ¹/₄ =

7. 7 ³/₇ =

8. ¹⁴⁰/₁₀₀ =

9. ⁵⁶⁰⁸/₈ =

10. ⁶⁴²⁶/₁₀ =

Continued on next page

Continued from previous page

Change the following decimals to fractions:

11. .25 =

12. .625 =

13. .300 =

14. .03125 =

15. .8125 =

16. 2.125 =

17. 24.375 =

18. .09375 =

19. .40625 =

20. .015625 =

ADDITION AND SUBTRACTION OF DECIMALS

When adding or subtracting decimals, write the figures to be added so that the decimal points are in straight vertical columns. Once the numbers are arranged in this manner, proceed as when adding or subtracting whole numbers.

EXAMPLE: Add .25 and 1.486

SOLUTION:

```
  .25     You can add a zero after the 5 without          .250
+ 1.486   changing the value of the decimal.            + 1.486
                                                          1.736
```

EXAMPLE: Subtract .24 from 1.527

SOLUTION:

```
  1.527
+  .240  ← Add a zero here to make the equation
  1.287    neater and easier to solve.
```

Errors in adding or subtracting decimals may be reduced by adding zeros to some of the numbers to act as place holders and to fill the empty spaces in the columns.

EXAMPLE:

Add .3, 1.25, .085 and 12.1

SOLUTION:

Aligning decimalsAligning decimals and adding zeros

```
      .3              .300
     1.25            1.250
      .085            .085
   +12.1           +12.100
    13.735          13.735
```

PRACTICE SET #8

Add or subtract the following decimal quantities:

1. .2 + .5 =

2. 2.2 + 24.3 =

3. 642.1 + 23.002 + .0789 =

4. 1.3 − .07 =

5. .006 − .00010 =

6. 5.726 − .3 =

7. 27.3 + 17.8 − 2.66 + .321 =

8. 1567.31 + .0013 − 17 =

9. A contractor's records show that a job's materials cost was $1289.45, the labor cost $972.68, and the overhead was $89.66. If he makes a profit of $368.00, what was the total job cost to the owner?

10. What is the outside pipe diameter if the metal thickness is .280" & the inside diameter is 8.065"?

11. What is the total thickness of shims: 0.007", 0.130", 0.125", .008" and .187"?

12. If the following amperes exist in the individual currents, find the total current: 0.216 amps, 1.67 amps, 2.0 amps and 1.837 amps?

13. A contractor purchases the following materials: 1400 feet of BX cable for $.03 per foot, 205 feet of 1" conduit at $.15 per foot, and 25 shallow boxes at $.73 each. What is the total material cost?

14. A plumber worked five days on a residential job. The number of hours worked were 10.75, 7.5, 9.5, 8.0 and 8.5. How many hours over 40 did he work?

15. A 20 foot piece of pipe had the following pieces cut from it: 1.25', 6.33', 2.625', 5.835' and 2.125'. How long was the remaining piece of pipe? *(Disregard cutting waste)*

MULTIPLICATION OF DECIMALS

To multiply decimals, proceed as when multiplying whole numbers. But in the product (answer), beginning at the right side, count off as many decimal places as there are on the right side of the decimal points in the numbers being multiplied.

EXAMPLE:

Multiply 2.62 by 2.6.

SOLUTION:

```
   2.62
x  2.6
   1572
   524
   6.812
```

→ Three numbers to the right of decimals in numbers being multiplied means the decimal moves three places to the left of the product.

We note that there are two decimal places to the right of the decimal in 2.62 and one decimal place to the right of the decimal in 2.6 for a total of three decimal places.

Beginning with the right hand number in the answer, count back a total of three places:

$$6 \quad 8 \quad 1 \quad 2 = 6.812$$

three two one start

EXAMPLE:

Multiply .214 by .212.

SOLUTION:

```
    .214
x   .212
    428
    214
+   428
    45368
```

As there are six digits to the right of the decimal points in the numbers being multiplied, we start at the 8 and move six decimal points to the left, needing to add a zero to achieve all six decimal spaces, resulting in our answer = .045368

Since there are six decimal places in the two numbers being multiplied and only five numbers, we must add a zero to fill the empty sixth place. Decimal places in the numbers being multiplied must have the same number of decimals in product or answer.

MULTIPLICATION OF DECIMALS

To divide decimals, we proceed as when dividing whole numbers, then we position the decimal point properly.

DIVIDING DECIMALS BY WHOLE NUMBERS

When a decimal is divided by a whole number, the decimal point in the answer is placed directly above the decimal point in the number being divided.

EXAMPLE: Divide .84 by 2

SOLUTION:

$$
\begin{array}{r}
.42 \\
2\overline{)\,.84} = .42 \\
\underline{8} \\
4 \\
\underline{4} \\
0
\end{array}
$$

DIVIDING WHOLE NUMBERS BY DECIMALS

When dividing a whole number by a decimal, move the decimal point of the divisor to the right until it becomes a whole number. Move the decimal point in the number being divided the same number of places to the right, adding zeros if needed. Then proceed as when dividing decimals by whole numbers.

EXAMPLE: Divide 72 by .3

SOLUTION:

$$
.3\overline{)\,72} = 3\overline{)\,720.} = 3\overline{)\,720.} = 240
$$

$$
\begin{array}{r}
240. \\
3\overline{)\,720} \\
\underline{6} \\
12 \\
\underline{12} \\
00
\end{array}
$$

Use the symbol "∧" to mark the new position of the decimal point. The important fact to remember in division with decimals is to make the divisor (the number outside the division sign) a whole number by moving the decimal point to the right. You must, of course, move the decimal the same number of places in the dividend (under the division sign). If there is a remainder, add as many zeros as you want to the number being divided and continue the division.

EXAMPLE:

Divide .296 by .25

SOLUTION:

```
         1.184
.25ʌ /.29ʌ600    (two zeros added to continue the division)
      25
      ‾‾
      46
      25
      ‾‾
      210
      200
      ‾‾‾
       100
       100
       ‾‾‾
```

DECIMAL MULTIPLICATION AND DIVISION WITH TENS

To multiply a decimal or a whole number by 10 or any multiple of 10, move the decimal point as many places to the <u>right</u> as there are zeros in the multiplier.

EXAMPLES

25 x 10 = 250 (one zero, so move the decimal one place to the right)

.51 x 10 = 5.1 (one zero, so move the decimal one place to the right)

.0867 x 1000 = 86.7 (three zeros, so move the decimal three places)

29.6238 x 100,000 = 29.6238 (five zeros, so move the decimal five places to the right)

To divide a decimal or a whole number by ten or any multiple of ten, move the decimal point to the <u>left</u> as many places as there are zeros in the divisor.

EXAMPLES

25 ÷ 10 = 2.5

7.6 ÷ 10 = .76

.0793 ÷ 100 = .000793

423.6 ÷ 10,000 = .04236

PRACTICE SET #9

Multiply or divide the following decimal quantities:

1. 1.7 x 4 =
2. 19.68 x 3.2 =
3. .08 x .325 =
4. 567.1 x .001 x .86 =
5. 45 x 100.1 =
6. .00063 x 100,000 =
7. 684.6 ÷ 10 =
8. 6846 ÷ 100 =
9. 2.068 ÷ 32 =
10. .068842 ÷ 46 =
11. 89,342 ÷ .006 =
12. .015736 ÷ 5.322 =
13. A contractor removed 385.9 cu yds. of earth from a construction site. If a truck can haul 1.7 cu. yds. per load, how many truck loads were removed?
14. 1456 machine bolts weigh 586.5 pounds. How much does one bolt weigh?
15. How much thicker are 22 sheets of .375" thick plywood than 25 sheets .250" thick?
16. Find the total weight of a full 52 gallon steel drum and its contents if one gallon of tar weighs 8.03 lbs. and the drum itself weighs 37.5 lbs.
17. What is the tax bill for a contractor if his shop and tools have an assessed valuation of $5500.00 and the tax rate is 22 $\frac{1}{2}$ mils (1 mil = .001)?
18. Working 7.25 hours per day, how many days would be required to do a job taking 83.375 hours?
19. A completed house costs $125,860. What is the average cost per square foot if the house contained 1520 square feet?
20. A platform 151.25 inches wide is to be covered with 2 x 6 (each 5.25 inches in width). How many pieces of 2 x 6 will be needed?

ROUNDING OFF DECIMALS

The degree of precision to which a quantity is to be calculated determines how accurate an answer to a problem is to be expressed. For instance, if you were gapping a spark plug, accuracy to the nearest .001" might be desirable. However, if you are measuring the distance between New York City and Miami, an answer to the nearest mile or even 100 miles might be adequate.

If you are taking a competancy test and the answers are given in multiple choice form, look at those answers to see how close together they are. If a brick quantity take off problem has four answer selections of 950, 960, 970 and 975 brick, you must proceed very carefully. However, answers of 800, 975, 1090 and 1200 will allow you to work more quickly and with less accuracy.

The process of expressing a number to a desired degree of accuracy may call for rounding off of that number. To refresh your memory, the rounding off process is as follows:

EXAMPLE:

Round off the decimal .749742 to three places or the nearest .001.

SOLUTION:

Locating the third place digit, which is "9" in this example, we look at the following digit, a "7". Since this 7 is greater than 5, add 1 to 9 giving .749 + .001 = .750.

*In other words, locate the digit in the place which indicates the required degree of precision.

*Increase that digit by (1) if the digit which immediately follows is (5) or more and drop the "(5) or more" digit and all digits that follow.

*Leave the digit as is if the digit which follows is less than (5) and drop all digits that follow.

Rounding off the following numbers to the nearest ten:

473 rounds off to 470

97 rounds off to 100

915 rounds off to 920

4376 rounds off to 4380

Rounding off the following numbers to the nearest one:

19.3 rounds off to 19

27.9 rounds off to 28

587.50 rounds off to 588

7436.49 rounds off to 7436

Rounding off the following decimals to the nearest hundredth
(or two places to the right of the decimal):

560.7921 rounds off to 560.79

21.3867 rounds off to 21.39

102.457 rounds off to 102.46

98.05005 rounds off to 98.05

With practice, rounding off answers becomes a simple process that can be done just by looking at an answer.

❸ *Inches-Feet Conversions*

INCHES TO FEET

If you are, say, finding the volume of a concrete footing with dimensions of 6" x 10" x 50', you must either change all dimensions to inches or to feet before multiplying. In this problem, it is probably best to convert all dimensions to foot-equivalents to keep the numbers small and to save time.

To change inches to a foot-equivalent, divide the number of inches by 12.

EXAMPLE:
Find the volume of a footing 6" x 10" x 50'

SOLUTION:

Decimal equivalents

$$12\overline{)6.00} = .5'$$

```
      .50
12 / 6.00  = .5'
    6 0
    ----
      00
      00
    ----
```

```
       .833
12 / 10.000  = .83' (approx.)
     96
    ----
     40
     36
    ----
      4
```

Fractional Equivalents

$$\frac{6"}{12} = \frac{1'}{2}$$

$$\frac{10"}{12} = \frac{5'}{6}$$

In other words, 6 inches equals .5 feet equals $\frac{1}{2}$ feet. You may find it easier to work with decimals, or the fractional expression may be best.

To find the volume of the concrete footing with dimensions of 6" x 10" x 50' multiply the depth by the width by the length (problems on volume are covered in more detail in a later chapter).

The solution by utilizing both decimals or fractions appears at the top of the next page.

Feet (Decimals)
.5' x .83' x 50' = 20.75 cubic feet

Feet (Fractions)

$$\frac{1}{\underset{1}{\cancel{2}}} \times \frac{5}{6} \times \overset{25}{\cancel{50}} = \frac{125}{6} = 20\,^{5}/_{6} \text{ cubic feet}$$

The decimal answer above is slightly more inaccurate due to the rounding off of .83333... to .83. The answer is, however, close enough for estimating or test purposes.

*To change inches to feet where a fractional part of an inch is involved,
first change inches to decimal inches, then divide by 12.*

EXAMPLE:

Change 4 $^{3}/_{8}$" to feet.

SOLUTION:

$$4\,^{3}/_{8} = {}^{35}/_{8}" = 8\overline{)35.00} = 4.375" \quad \text{and} \quad 12\overline{)4.37500} = .3645' \text{ or rounded off to } .36'$$

```
      4.375                      .3645
 8 ) 35.00              12 ) 4.37500
     32                      3 6
     ──                      ──
     30                       77
     24                       72
     ──                       ──
     60                       55
     56                       48
     ──                       ──
     40                       70
     40                       60
     ──                       ───
                             100
```

EXAMPLE:

Change 16 $^{3}/_{4}$" to feet.

SOLUTION:

16 $^{3}/_{4}$" equals one foot and 4 $^{3}/_{4}$ inches. The 4 $^{3}/_{4}$ inches only must be changed to feet, since the additional one foot is already in terms of feet. Thus,

$$4\,^{3}/_{4}" = {}^{19}/_{4}" = 4\overline{)19.00} = 4.75" \quad \text{and} \quad 12\overline{)4.7500} = .3958'$$

```
      4.75                      .3958
 4 ) 19.00             12 ) 4.7500
     16                      3 6
     ──                      ──
     3 0                     1 15
     2 8                     1 08
     ──                      ───
      20                      70
      20                      60
      ──                      ──
                             100
                              96
                             ──
                              4
```

So, 1.0' + .3958' = 1.3658', rounded off to 1.396'.
Therefore, 16 $^{3}/_{4}$" ≈ 1.396'

FEET TO INCHES

To change feet to inches, multiply feet by 12.

EXAMPLE:

Express 5 feet in terms of inches.

SOLUTION:

5 x 12 = 60"

EXAMPLE:

Express 8.25 feet as inches.

SOLUTION:

8.25 x 12 = 99 inches

EXAMPLE:

Express 10 $^1/_3$ feet as inches.

SOLUTION:

$10 \ ^1/_3 = \ ^{31}/_3$, so $\dfrac{31}{\cancel{3}} \times \cancel{12}^{\,4} = 124$ inches

PRACTICE SET #10

Change the following to feet:

1. 4" =
2. 9" =
3. 4 $^1/_2$" =
4. 8 $^7/_8$" =
5. 17 $^5/_{16}$" =
6. 256 $^7/_{32}$" =
7. 5 $^{31}/_{32}$" =
8. $^{31}/_{32}$" =

Change the following to inches:

9. 10 feet =
10. 5.25 feet =
11. 26 $^1/_4$ feet =
12. $^2/_3$ foot =
13. 8.625 feet =
14. 120.5 feet =
15. $^1/_{10}$ foot =

ADDITION AND SUBTRACTION OF INCHES AND FEET

Since inches and feet can be expressed in several different forms (such as 6' 8", 6 $^2/_3$' or 6.67'), adding and subtracting inches, feet, yards, miles, etc. can be done in several ways. Let's illustrate with an example.

EXAMPLE:

Add 15' 9" and 16' 8".

SOLUTION:

1) Conversion to inches:
   15' 9" = (15' x 12) + 9" = 180" + 9" = 189"
   16' 8" = (16' x 12) + 8" = 192" + 8" = 200"
   189" + 200" = 389"
   389" ÷ 12 inches/foot = 32' 5"

2) Conversion to feet:
   $$15' \, 9" = \quad 15\,^9/_{12}'$$
   $$16' \, 8" = +\, 16\,^8/_{12}'$$
   $$\overline{\qquad 31\,^{17}/_{12}'} = 32\,^5/_{12}'$$

3) Conversion to decimal feet:
   $$15' \, 9" = \quad 15.75'$$
   $$16' \, 8" = +\, 16.67'$$
   $$\overline{\qquad 32.42'} = 32.42'$$

4) Adding feet and adding inches and combining:
   $$15' \, 9"$$
   $$+\, 16' \, 8"$$
   $$\overline{\quad 31' \, 17"} = 31' + 1' \, 5" = 32' \, 5"$$

Any one of the above methods will work for addition or subtraction. Use the procedure that seems most convenient to you. If you choose #4 above, it may be necessary to "borrow" 12" as follows:

EXAMPLE:

Subtract 15' 9" from 38' 8".

SOLUTION:

$$38' \, 8" = 37' \, 8" + 12" = \quad 37' \, 20"$$
$$-\, 15' \, 9" = \qquad\qquad\qquad -\, 15' \quad 9"$$
$$\overline{\qquad\qquad\qquad\qquad\qquad 22' \, 11"}$$

PRACTICE SET #11

Add or Subtract the following measurements:

1. 5' 6" + 7' 4" =

2. 8' 9" + 12' 5" =

3. 10' 5 $\frac{3}{4}$" + 2' 7 $\frac{3}{4}$" + 6' 9 $\frac{7}{8}$" =

4. 12' 4 $\frac{1}{2}$" − 5' 7 $\frac{3}{8}$" =

5. 10' 0" − 5' 7 $\frac{3}{8}$" =

6. 22' 11 $\frac{7}{8}$" + 12' 9 $\frac{5}{8}$" + 2' 10 $\frac{15}{16}$" − 1' 6" =

7. 50' 0" − 5' 3" − 8' 7 $\frac{1}{2}$" − 2' 11 $\frac{3}{4}$" − 8' 3 $\frac{3}{4}$" =

8. $\frac{1}{2}$" + 7' 3 $\frac{3}{8}$" + 2' 9 $\frac{7}{8}$" − 10' 0" =

9. 17.25' − 8' 4" =

10. 2.125' + 7' 1" + 9' 11" − 10.5' =

11. 100.00' + 7 .635' + 5 $\frac{3}{8}$" − 1" + 1.0' =

12. .859375' + 5' 0" + $\frac{9}{32}$" =

MULTIPLICATION AND DIVISION OF INCHES AND FEET

When finding areas or volumes, multiplication and division of inches and feet become necessary.
As explained for adding and subtracting above, several methods may be used to obtain a correct answer.

EXAMPLE:

Multiply 15' 9" by 16' 8".

SOLUTION:

1) Conversion to inches:
   15' 9" = 189"
   16' 8" = 200"
   189" x 200" = 37,800 sq. in. ÷ 144 sq. in./sq. ft. = 262.5 sq. ft.

This method is generally not desirable since the numbers involved in the calculations tend to become very large.

2) Conversion to fractional feet:

15' 9" = 15 $^3/_4$' = $^{63}/_4$'

16' 8" = 16 $^2/_3$' = $^{50}/_3$'

63/4 x 50/3 = 3150/12 = 262.5 sq. ft.

3) Conversion to decimal feet:

15' 9" = 15.75'

16' 8" = 16.67'

15.75 x 16.67 = 262.5 sq. ft.

4) Procedure #4 for addition will not work in multiplication or division.

EXAMPLE:

Divide 40' 8" by 6.

SOLUTION:

1) Conversion to inches

40' 8" = (40 x 12) + 8" = 480 + 8 = 488"

488" ÷ 6 = 81 $^2/_6$" = 81 $^1/_3$" = 6' 9 $^1/_3$"

2) Conversion to fractional feet:

40' 8" = 40 $^2/_3$' = $^{122}/_3$'

$$\frac{122}{3} \div 6 = \frac{122}{3} \times \frac{1}{6} = \frac{122}{18} = 6 \ ^{14}/_{18}' = 6 \ ^7/_9' = 6.78'$$

3) Conversion to decimal feet:

40' 8" = 40.67'

40.67' ÷ 6 = 6.78'

Again, choose the method you like best. Generally, conversion to fractional or decimal feet will work best when multiplying or dividing. Working with practice problems using each of the methods will help you see which you will want to use.

The procedures given in this chapter are for inches and feet, since these units are encountered most often. The principles hold true when performing operations involving any unlike units such as inches-feet, feet-yards, hours-days, cubic inches-gallons, and so on. In all cases, express the quantities in some common unit, then proceed to work with the numbers.

Some of the more common conversion factors are given in Table 1, page 81, in the back of this manual.

EXAMPLE:

How many hours are in 1 week, 2 days and 4 hours?

SOLUTION:

1 week = 7 days x 24 hours/day = 168 hours
2 days = 2 days x 24 hours/day = 48 hours
4 hours =   <u> 4 hours</u>
                                      220 hours TOTAL

EXAMPLE:

How many gallons are in 3 cubic feet and 462 cubic inches? Note that there are 7.5 gallons in a cubic foot of water.

SOLUTION:

$$\text{3 cubic feet x 7.5 gal/cu.ft.} \quad = \quad \text{22.5 gallons}$$

$$\text{462 cubic inches} \div \text{1728 cu. in./cu. ft.} = .267 \text{ cu.ft.}$$

$$\text{.267 cubic feet x 7.5 gal/cu.ft} = \quad \underline{+ \text{ 2.0 gallons}}$$

$$\text{24.5 gallons TOTAL}$$

AT A GLANCE

You can cancel units like simplifying fractions to properly display your answers in the correct units. In the example above,

3 cubic feet x 7.5 <u>gallons</u>   becomes 3 ~~cubic feet~~ x 7.5 <u>gallons</u>
              cubic feet   *~~cubic feet~~*

and results in 22.5 gallons as the correct answer and unit label.

PRACTICE SET #12

Multiply or divide the following measurements:

1. 2' 0" x 5' 0" =

2. 7' 5" x 6' 6" =

3. 8 $\frac{1}{2}$' x 4' 8" =

4. 10.25' x 3 $\frac{7}{12}$' =

5. $\frac{3}{4}$' x 8" x 10.5' =

6. 16" x 22" x 20 $\frac{3}{4}$' =

7. 10' 0" ÷ 2' 6" =

8. 5 $\frac{2}{3}$' ÷ 1 $\frac{1}{5}$' =

9. 16' 8" ÷ 9" =

10. 24.25' ÷ $\frac{1}{2}$" =

11. 120.5' ÷ 5' $\frac{1}{4}$" =

12. 88' 6" ÷ 18' 4" =

13. A brick wall is to be 28' 2 $\frac{1}{4}$" long. How many brick per course will there be if one brick and mortar joint combined length is 8 $\frac{1}{4}$"?

14. How many pieces of pipe each 8 $\frac{3}{4}$" long can be cut from a 20' length? *Neglect cut waste*

15. A strip of 1 x 2 lumber is cut into 10 $\frac{5}{16}$ inch lengths for shelf cleats. If $\frac{1}{8}$" is lost with each cut, how much length is used to make 12 cleats *(assume 12 cuts)*?

16. A stairs is to be 8' 5 $\frac{1}{2}$" high. How many risers will be built is each is to be 7 $\frac{1}{4}$" high?

17. A room is built that is 384" long. If joists are spaced 16" on center, how many joists are needed *(assume double joists each end, single joists elsewhere)*?

18. Reinforcing steel is placed in a retaining wall that is 154' long. If the effective length of a steel bar is 19.25', and if 2 pieces of steel will be laid side by side, how many pieces will be required?

19. If a utility pole is to be installed every 125' 6", how many poles are needed for a one mile stretch *(5,280' per mile)*?

20. 177' 10" of pipe was used on a job. If each piece was 8.0833' long, how many pieces were used?

4 Percentages

Percentage is a term used in math, business and our everyday lives to indicate a number of parts of one hundred. One percent (1%) of a quantity means $^{1}/_{100}$ (fraction form) or .01 (decimal form). Our most typical use is money when we refer to coins. A quarter is .25 cents or a "quarter" of a dollar, or 25% of a dollar.

Percents may be added, subtracted, multiplied or divided.:

5% + 8% = 13%

10% − 4% = 6%

3% x 4 = 12%

25% ÷ 5% = 5

CHANGING PERCENT TO DECIMALS

Percents may be changed to decimals by moving the decimal point two places to the left and dropping the % sign.

EXAMPLE:

6% = .06

42% = .42

85% = .85

AT A GLANCE

It is important to drop the % sign as in the example to the left. 6% is __NOT__ .06%

CHANGING PERCENT TO COMMON FRACTIONS

To change a percent to a common fraction, change the % to a decimal equivalent, then change the decimal to a fraction.

EXAMPLE:

5% = .05 = 5/100 = $^{1}/_{20}$

25% = .25 = 25/100 = $^{1}/_{4}$

80% = .80 = 80/100 = $^{4}/_{5}$

.1% = .001 = $^{1}/_{1000}$

CHANGING A DECIMAL TO A PERCENT

*To change a decimal to a percent, move the decimal point
two places to the right and add the % sign.*

EXAMPLE:

.12 = 12%

.55 = 55%

1.9 = 190%

CHANGING COMMON FRACTIONS TO PERCENT

*The first step is to change the common fraction to a decimal by dividing
the numerator by the denominator. Then change the decimal to a percent.*

EXAMPLE:

$^1/_5$ = .20 = 20%

$^1/_8$ = .125 = 12.5%

$^{13}/_{16}$ = .8125 = 81.25%

THE THREE TYPES OF PERCENTAGE PROBLEMS

Understanding some basic equations and how the numbers interrelate is a basic skill needed to pass exams. Ohm's law, Pythagorean's Theorem, even the famous formula for relativity are essentially the same — a specific relationship of numbers. Each component of the formula is needed to solve for your answer — or more specifically, the recipe or formula can be used to determine one of the missing pieces of data (the answer). While we will not be pursuing calculus or the theory of relativity here, beginning to understand the relationships of numbers will help greatly in your field.

Practically, when determining percentages we have three main "ingredients". The first is the "base", the number we are going to be finding some percent of. The second is the "rate percent", the fractional amount of the base we want to determine. The final piece is the "percentage", the fractional result of the base when multiplied by the percent rate.

Simple interest, volumes and capacities are all examples of percentages diagnosed as:

$$\%\ =\ B\ x\ R$$

Percentage   Base   Rate

1) Given the base and the rate, find the percentage.

EXAMPLE: Find 12% of 200.

SOLUTION:

12% = .12, so 200 x .12 = 24.00

EXAMPLE: Find 20% of 75.

SOLUTION:

Formula becomes % = 75 x .20

             % = 15

                15 is 20% of 75

(Percentage = Base x Percent Rate or P = B x R)

2) Given the base and the percentage, find what percent one number is of another.

EXAMPLE: 10 is what % of 50?

SOLUTION:

$\dfrac{10}{50} = \dfrac{1}{5} = 20\%$ (dividing 5 into 1)

EXAMPLE: 20 is what % of 5?

SOLUTION:

$\dfrac{20}{5} = \dfrac{4}{1} = 400\%$

(Percentage Rate = Percentage ÷ Base or R = P/B)

3) <u>Given the rate and percentage, find the base when a percent of that base is known.</u>

EXAMPLE: 40 is 25% of what amount?

SOLUTION:

25% = 25/100 = 1/4, so 40 ÷ 1/4 = 40 x 4/1 = 160
25% = percent rate
  40 = percentage
 160 = base

EXAMPLE: 60 is 20% of what amount?

SOLUTION:

20% = 20/100 = 1/5, so 1/5 = 60 x 5/1 = 300
60 is 20% of 300

> **(Base = Percentage ÷ Percent Rate or B = P/R)**

The formulas given above help make solving percentage problems easier if you learn to identify the base, the percent rate and the percentage.

PRACTICE SET #13

Change the following to percent:

1. .5 =

2. .75 =

3. 1.30 =

4. .03 =

5. .002 =

6. $^1/_4$ =

7. $^1/_{10}$ =

8. $^1/_{50}$ =

9. $^5/_8$ =

10. $^9/_5$ =

Continued on next page

11. $^{17}/_{100} =$

12. $^{9}/_{200} =$

13. Find 20% of 600 gallons.

14. What is $^{1}/_{2}$ of 1% of 250 tons?

15. $^{1}/_{3}$ is what % of $^{2}/_{3}$?

16. 10 is 50% of what number?

17. 25% of a price equals $4800. What is the price?

18. 9¢ is what % of 36¢?

19. What number increased by 20% of itself equals 60?

20. $^{1}/_{4}$ is what % of $^{1}/_{3}$?

21. What is 140% of $200?

22. 4 = 15% of what number?

23. A concrete batch contains 8% cement by weight. If the total batch weight is 10,000 pounds, how many pounds of cement are there?

24. A contractor was assessed a 3% penalty for not completing a job on time. If the contract price was $68,000, what was the penalty?

25. A brick layer wasted 2.8% of the brick while laying a brick wall. If 28 bricks were wasted, how many bricks were in the wall?

26. Find the amount of board feet of lumber to be ordered if 2500 board feet are needed and 10% is allowed for waste.

27. The landscaping for a $480,000 home was $6,000. What percent of the total cost is this?

28. To his estimate of $50,000 the contractor adds 2% for incidentals and 8% of the estimate and incidentals for profit — what is his bid?

29. If 5.85% of wages are withheld from an employee on the first $10,800 for social security, what is the maximum amount that must be paid by that employee?

30. A bank pays 5% interest for money left in savings accounts. If the interest earned for the year was $22.00, what was the amount of savings?

5 Ratio

A <u>ratio</u> is the relation between two numbers or values. The ratio may be written as a fraction, 2/3; as a division, 2 ÷ 3; or with the ratio sign (:) 2:3. When the ratio is used, it is read "2 to 3" or "2 is to 3". The numbers "2" and "3" are called <u>terms</u>.

It is important to understand that the ratio doesn't reflect a total of the terms, but instead represents a relationship between the terms. You can use this ratio to determine totals for a given task at hand.

Both terms of a ratio may be multiplied or divided by the same number without changing the value of the ratio. Thus, 3:6 = 6:12 (multiplying both terms by 2) or 3:6 = 1:2 (dividing both terms by 3).

> *To separate a quantity according to a given ratio, add the terms to find the total number of parts. Express each term as a fractional part of the whole and multiply by the whole part.*

A typical ratio problem might be as follows:

EXAMPLE:

If a mixture requires 1 part of alcohol to 3 parts of water, how many gallons of alcohol and gallons of water will be needed for a 1000 gallon mixture?

SOLUTION:

1) Adding the terms, 1 + 3 = 4.

2) Expressing each term as a fractional part of the whole gives $\frac{1}{4}$ and $\frac{3}{4}$.

3) Multiplying by the whole part =

  1/4 x 1000 = 250 gallons alcohol

  3/4 x 1000 = 750 gallons water

Ratios do not have to have only two terms. The relationship or "recipe" can include any number of terms.

EXAMPLE:

1600 gallons of diesel fuel have to be distributed to three pieces of equipment in the ratio 8:11:13. How many gallons should each receive?

SOLUTION:

1) 8 + 11 + 13 = 32

2) 8/32, 11/32 and 13/32 are the fractional parts

3) 8/32 x 1600 = 400 gallons

    1/32 x 1600 = 550 gallons

    13/32 x 1600 = 650 gallons

ROOF PITCH

Since the pitch of a roof is expressed as a ratio, a discussion of roof pitches is appropriate at this time.

The pitch of a roof is the angle or slope of the roof from the plate to the ridge. It is the ratio of the rise to the span.

The rise of a roof is the vertical distance that the rafters or trusses climb from the eaves to the ridge.

The run is the horizontal distance which the rafters or trusses cover from the eaves to the ridge.

The span is equal to twice the run. It is also equal to the rise divided by the pitch.

$$Pitch = \frac{rise}{span} = \frac{rise}{2 \text{ x } run}$$

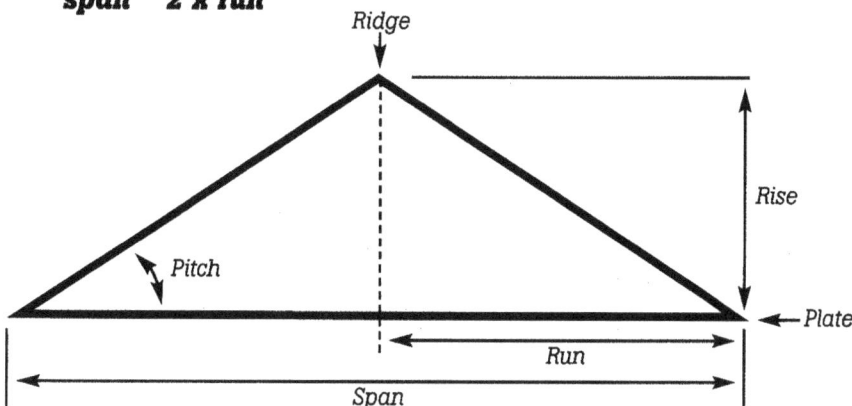

EXAMPLE:

What is the pitch of a roof if the span is 24' and the rise is 4'?

SOLUTION:

$$\text{Pitch} = \frac{\text{rise}}{\text{span}} = \frac{4}{24} = \frac{1}{6}$$

The word "pitch" is commonly used in error by many tradesmen on construction jobs to mean the ratio of rise to run. In other words, you may hear that a certain roof "has a pitch of 3 in 12", meaning, a rise of 3" for each 12" of horizontal run. THIS IS INCORRECT. The correct term for the amount of rise for every twelve inches of run is the slope.

Pitch is the ratio of the rise to the total span, where the span is twice the run. A 3 in 12 roof has a pitch of 1/8 ($\frac{3}{2 \times 12} = \frac{3}{24} = \frac{1}{8}$).

PRACTICE SET #14

Find the ratio of the first quantity to the second:

1. 8 hours to 1 day   3. 9 months to 5 years   5. 10 pounds to 1 ton

2. 10 ozs. to 1 lb.   4. 528 feet to 2 miles   6. 80¢ to $10.00

7. Pitch is defined as "the ratio of rise to span". If the rise of a roof is 8' and the span is 24', what is the pitch?

8. If the rise is 12' and the run is 12', what is the pitch?

9. A gabled roof rises 14' and spans 28'. What is the pitch?

10. $5000 is to be distributed among three subcontractors in the ratio of 6 to 3 to 1. How much will each receive?

11. Bronze consists of 6 parts tin to 19 parts copper. How many pounds of copper are in a 500 lb. bronze statue?

12. What is the ratio of number of primary turns to secondary turns if the primary transformer has 45 turns and the secondary transformer has 500 turns?

13. 800 drums of diesel fuel are to be distributed to 4 construction jobs in the ratio of 10 to 5 to 3 to 2. How many drums of fuel will each job receive?

14. A fuel mixture consists of 2 parts oil to 83 parts gasoline. How many quarts of oil would be in a 340 gallon mixture?

6 *Proportions*

A <u>proportion</u> is an expression of equality between two ratios. It may be written with the double colon or proportion sign (::) or with an equal sign (=).

The proportion 1 : 2 :: 4 : 8 is read "one is to two as four is to eight" or "$\frac{1}{2}$ equals $\frac{4}{8}$".

The numbers "1" and "8", or the first and last numbers, are called the <u>extremes</u> and the numbers "2" and "4", or the second and third numbers, are called the <u>means</u>.

The product of the extremes must equal the product of the means for a proportion to be valid.

In the example above,

        Extremes: 1 x 8 = 8
        Means: 2 x 4 = 8
                         8 = 8 OK

When written as $\frac{1}{2} \diagup\!\!\!\!\diagdown \frac{4}{8}$ the means and the extremes are diagonally opposite each other.

If we know three of the terms of a proportion, we can find the fourth term by the following method:

    1) Let "X" stand for the unknown term.
    2) Multiply the extremes.
    3) Multiply the means.
    4) Set the two products equal to each other and solve for X.

EXAMPLE:

Find the value of the missing term.

    2 : 5 :: 4 : X

SOLUTION:

    Extremes = 2 and X
    Means = 5 and 4
    Multiplying: $2X = 5 \times 4$
                    $2X = 20$
                     $X = 20/2 = 10$

    Thus, 2 : 5 :: 4 : 10 or $^2/_5 = {}^4/_{10}$
    (Checking for correctness, $2 \times 10 = 4 \times 5 = 20$) OK

EXAMPLE:

Solve for the missing term where $X/9 = 4/36$

SOLUTION:

$36X = 9 \times 4$; $36X = 36$; $X = 1$
Thus, $\dfrac{1}{9} = \dfrac{4}{36}$

EXAMPLE:

Solve for X where 2 : 8 :: 10 : X

SOLUTION:

$2X = 8 \times 10$; $2X = 80$; $X = 40$
Thus, $\dfrac{2}{8} = \dfrac{10}{40}$

The proportions above have been solved using the equation method of solving problems with an unknown. This process is covered in more detail in the next two chapters.

PRACTICE SET #15

Find the missing term:

1. 2 : 3 :: 4 : ? =

2. 5 : 10 :: 40 : ? =

3. 2 : ? :: 1 : 2 =

4. 3 : ? :: 4 : 12 =

5. 4 : 24 :: ? : 6 =

6. 8 : 18 :: ? : 72 =

7. 3 : 10 :: 60 : ? =

8. ? : 5 :: 32 : 20 =

9. ? : 2 :: 40 : 80 =

10. 3 : ? :: 21 : 49 =

Solve the following proportion problems:

11. If 1 gallon of paint is needed to prime 300 sq. ft. of plaster, how many gallons are needed for 2400 sq. ft?

12. If 50 pounds of nails cost $6.00, how much does 200 pounds cost?

13. A cylindrical tank that is 10' deep holds 2500 gallons of water when full. How many gallons does it hold if the water depth is 1 foot?

14. If 1 square foot of wall contains 13 bricks, how many bricks are in 2005 sq. ft. of wall?

15. An earth embankment rises $\frac{1}{2}$ foot for every foot of level ground. How much will the embankment rise in 26 feet of level ground?

16. A copper wire 600 feet long has a resistance of 1.72 ohms. How long is a copper wire of the same area that has a resistance of 4.62 ohms?

17. A certain type of plaster work requires 1 $\frac{1}{2}$ cubic yards of sand per 100 square yards of work. How much sand is needed for 4200 square feet?

18. If a pole 18' high casts a shadow 20' long, how long a shadow would a pole 27' high cast?

19. A certain type of fitting runs 6 $\frac{1}{2}$ to the pound. How many fittings would be in 25 $\frac{1}{4}$ pounds?

20. For a house that is 51' wide and 77' long, a 20' wide addition is to be built having the same proportion as the house itself. How long will the addition be?

21. The scale on a drawing is 1" = 100'. If a pipe measures 6 $\frac{1}{2}$" on the drawing, how long is the pipe?

22. The scale on a map is 1" = 50,000 feet. How many miles are represented by a line that is 7.5" long?

⬢ 7 *Algebra Fundamentals*

Before studying powers, roots, formulas and equations, some basic discussion of algebraic theory is necessary.

In solving various problems in arithmetic, algebra, etc., it may become necessary to deal with negative numbers. These numbers have values less than zero and are written with a "−" sign. Positive numbers have values above zero and usually have the "+" sign. If no sign is indicated, the number is assumed to be positive.

Learning how to work with signed numbers helps form the basis for dealing with more complicated algebraic problems later.

ADDITION OF SIGNED NUMBERS

To add numbers of <u>like</u> signs, add the numbers and give the answer the sign that the numbers have.

EXAMPLE:

Add +12, +3 and +7

SOLUTION:

$$
\begin{array}{r}
+12 \\
+\ 3 \\
\underline{+\ 7} \\
+22
\end{array}
$$

EXAMPLE:

Add −2, −8 and −13

SOLUTION:

$$
\begin{array}{r}
-\ 2 \\
-\ 8 \\
\underline{-13} \\
-23
\end{array}
$$

To add numbers of <u>unlike</u> signs, group all positive and all negative numbers, subtract the smaller from the larger and give the answer the sign of the larger combination.

EXAMPLE:

Add −2, +7, −3, +14, +3 and −4.

SOLUTION:

$$+7 + 14 + 3 = +24$$
$$(-2) + (-3) + (-4) = -9$$
$$24 - 9 = +15$$

EXAMPLE:

Add +22, −43, −20 and +7.

SOLUTION:

$$+7 + 22 = +29$$
$$(-43) + (-20) = -63$$
$$29 - 63 = -34$$

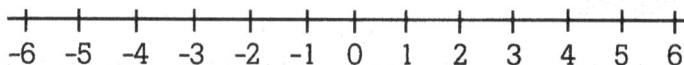

SUBTRACTION OF SIGNED NUMBERS

To subtract signed numbers, change the sign of the number to be subtracted (subtrahend) and proceed as when adding signed numbers.

EXAMPLE:

Subtract 10 from 23.

SOLUTION:

10 is the number being subtracted, so changing its sign and adding:
$$23 - 10 = 13$$

EXAMPLE:

Subtract −20 from −5.

SOLUTION:

Changing the sign of the number being subtracted (-5), we get
$$(-20) - (-5) = -20 + 5 = -15$$

MULTIPLICATION OF SIGNED NUMBERS

Multiplying any two numbers that have <u>like</u> signs will give a product that is positive. This is important as two negative numbers result in a positive product. Not exactly "two wrongs make a right", but an easy way to remember it.

EXAMPLE: Multiply +3 by +6.
SOLUTION:

3 x 6 = 18

EXAMPLE: Multiply -4 by -8.
SOLUTION:

(-4) x (-8) = +32

Multiplying any two numbers with <u>unlike</u> signs will give a product that is negative.

EXAMPLE: Multiply -5 by +6.
SOLUTION:

(-5) x 6 = -30

EXAMPLE: Multiply -3 by +4 by -8 by -2.
SOLUTION:

(-3) x 4 = -12;   (-12) x (-8) = +96;   96 x (-2) = -192

AT A GLANCE

It should be noted here that any odd number of negative digits will result in a negative product. Every two negative numbers multiplied together result in a positive product. This is a great way to check your answers' sign as shown in the second example to the left.

DIVISION OF SIGNED NUMBERS

If the two numbers in the division problem have like signs, the answer will be positive, the same as multiplication as detailed above.

EXAMPLE: Divide -12 by -2.
SOLUTION:

-12/-2 = +6

If the two numbers in the division problem have <u>unlike</u> signs, the answer will be negative.

EXAMPLE:

Divide –24 by +6.

SOLUTION:

–24/+6 = –4

PARENTHESES

Parentheses ( ) or brackets [ ] mean the quantities inside are to be grouped together and considered as a single quantity. Often, there is an arithmatic formula or directive inside the () or [].

12 – (8 + 2) is read "twelve minus the quantity eight plus two".

To solve problems containing parentheses, do the work within the parentheses first. Then remove the parentheses and continue.

EXAMPLE

5(6 + 3) = 5(9) = 45

Notice that taking 5 x 6 and adding 3 above would give 33, an incorrect answer. The operations within the parentheses must be carried out first.

EXAMPLE

$$\frac{30}{2(3 + 2)} = \frac{30}{2(5)} = \frac{30}{10} = 3$$

EXAMPLE

$$\frac{60 - (6 + 4)}{5(2 + 3)} = \frac{60 - (10)}{5(5)} = \frac{50}{25} = 2$$

EXAMPLE

$$\frac{(8 + 6) - (3 + 2)}{\frac{(9 - 6)}{12} \times 24} = \frac{(14) - (5)}{\frac{3}{12} \times 24} = \frac{9}{\frac{1}{4} \times 24} = \frac{9}{6} = \frac{3}{2} = 1\frac{1}{2}$$

Brackets are used to indicate that one or more quantites in parentheses are to be treated as one unit. Do the work within the parentheses first, then the brackets.

EXAMPLE

$24 \div [6(4 - 2)] = 24 \div [6(2)] = 24 \div [12] = 2$

EXAMPLE

$$\frac{3[6(2 + 5) + 7]}{7} = \frac{3[6(7) + 7]}{7} = \frac{3[42 + 7]}{7} = \frac{3[\cancel{49}]^{7}}{\cancel{7}} = 21$$

BASIC ALGEBRA

Now that we understand the rules used when working with signed numbers and parentheses, we can proceed to substituting letters or symbols for numbers. This is the most common way to find an unknown.

An expression in which letter symbols are used to represent numbers is called an algebraic expression. If you know the number values of the symbols, then you can find the numerical value of any algebraic equation.

The four arithmetical operations are:

"a + b" means that a and b are added together. If a = 3 and b = 4, then a + b = 7.

"a − b" means that b is subtracted from a. If a = 6 and b = 2, then a − b = 4.

"ab, a x b. a • b, (a)(b)" means that a is multiplied by b. Any one of the four ways shown may be used to indicate multiplication. If a = 5 and b = 7, then ab = 35.

"a ÷ b, a/b" means a is divided by b. If a = 10 and b = 5, then a/b = 2.

Here are some combinations of the four basic arithmetical operations (for our examples we will use the values a = 2, b = 3 and c = 4):

$$
\begin{aligned}
a + b + c &= \quad 2 + 3 + 4 = 9 \\
a - b + c &= \quad 2 - 3 + 4 = 3 \\
-a - b - c &= \quad -2 - 3 - 4 = -9 \\
a \times c &= \quad 2 \times 4 = 8 \\
a \div c &= \quad 2 \div 4 = \tfrac{1}{2} \\
\frac{a + c}{b} &= \quad \frac{2 + 4}{3} = \frac{6}{3} = 2 \\
\frac{a}{b} + \frac{c}{b} &= \quad \frac{2}{3} + \frac{4}{3} = \frac{6}{3} = 2 \\
\frac{c - a}{b} &= \quad \frac{4 - 2}{3} = 3\,\tfrac{1}{3}
\end{aligned}
$$

The use of letters and symbols become important when we want to describe rules and relationships in an abbreviated form. For instance, if we wish to state that

"Profit equals the margin minus the overhead"

we can assign letter symbols to this statement and write it in shorthand form as

$$P = M - O$$

where P = profit, M = margin and O = overhead.

The letter statement above is called an equation — a statement that two expressions are equal.

It should also be noted that the letter symbols can themselves represent a "sub-equation" whose answer fits into the main equation. In the example above, the letter "O" has many components such as rent, labor, taxes, insurance, etc. Solve for "O" first and plug the answer into the main profit equation to determine profitability as represented by $P = M - O$.

The following chapter will deal with equations and how to work and solve them.

PRACTICE SET #16

1. $+5 + 7 =$

2. $+10 - 5 + 2 - 6 =$

3. $14a + 6a =$

4. $6b - 2b =$

5. $+6 - (-5) =$

6. $8(4) + 3 =$

7. $(6 \div 2) + 7 =$

8. $5(a + 3) + 2 =$

9. $7a + 6b + 2a + 4b - 5b - (-4b) =$

10. $4(a + b) + 2a - b =$

11. $-24 \times 4 \div 8 =$

12. $(72 \div -24) + 6(-2 - 5) =$

13. $(2) \, (-3) \, (6) \, (4) \, (^1/_2) =$

Continued on next page

Continued from previous page

14. $\dfrac{(4)\ (5)}{(2)\ (5)} + \dfrac{(3)\ (-4)}{-10} =$

15. $-2[4(5 + 6)] =$

16. $3[2a(6 + 2)] + 7a(-2) =$

For the following algebraic expressions, find the numerical value if $a = 2$, $b = 3$ and $c = 4$:

17. $a + b + c =$

18. $a - b - c + a =$

19. $-(a + b + c) =$

20. $(a + b) - (a + b) =$

21. $a(b) + c =$

22. $a + b(c) =$

23. $1/2a + 1/3b + 1/4c =$

24. $\dfrac{(a + b)}{c} + \dfrac{(a + c)}{b} + \dfrac{(b + c)}{a} =$

25. $[a(b + c) + c] + b(a + c) =$

26. $\dfrac{a(b - c)}{a(b + c)} + b + c =$

27. $[-(abc)\ (abc)] =$

28. $\dfrac{2a + 3b + 4c}{29} - 1 =$

Basic Construction Math Review *3rd Edition*

53

⬡8 Equation Solving

A good understanding of the previous chapter on algebraic expressions will help you tremendously in learning how to solve equations.

As was mentioned in that chapter, an <u>equation</u> is a statement that two quantities or expressions are equal. Thus, common sense will tell us that

"What you do to one side of an equation must also be done to the other side to preserve equality."

We can add to, subtract from, multiply or divide both sides of the equation by the same quantity without changing the equality. Let's illustrate with equations.

ADDITION OF QUANTITIES

EXAMPLE:

$x - 4 = 8$, what does x equal?

SOLUTION:

$x - 4 + 4 = 8 + 4$ (we added a +4 to both sides of the equation so that the 4's
$x = 12$ on the left side would "cancel out", leaving us with x by itself)

In general, to solve an equation, we want all the unknown quantities (x, y, z, etc.) on one side of the equals sign and all the numbers on the other side. Then we can combine all unknowns as much as possible and combine all numbers to get a single number.

The general purpose of solving equations is to find an unknown — or the answer. The goal of performing identical mathematical operations to both sides of the equation should be to isolate the unknown. Once isolated it can be solved for rather easily.

This collecting of unknowns on one side of the equals sign is accomplished by adding, subtracting, multiplying and dividing by the quantities on the other.

Also, to check your answer, simply take the answer and substitute it back into the original equation. Both sides of the equation must produce the same number for the answer to be correct.

EXAMPLE:

$24 = x - 10$, solve for x and check answer.

SOLUTION:

$$24 = x - 10; \quad 24 + 10 = x - 10 + 10; \quad 34 = x$$

Check: $24 = (34) - 10; \quad 24 = 24$ $\checkmark$

SUBTRACTION OF QUANTITIES

EXAMPLE:

If $y + 5 = 12$, $y = ?$

SOLUTION:

$$y + 5 - 5 = 12 - 5; \quad y = 7$$

Check: $(7) + 5 = 12; \quad 12 = 12$ $\checkmark$

MULTIPLICATION OF QUANTITIES

EXAMPLE:

$X/4 = 5$

SOLUTION:

$\dfrac{X}{4} \times 4 = 5 \times 4$, multiplying both sides by 4

$\dfrac{X}{\cancel{4}} \times \cancel{4} = 20$, cancelling; $\quad x = 20$

Check: $20/4 = 5; \quad 5 = 5$ $\checkmark$

DIVISION OF QUANTITIES

EXAMPLE:

$3a = 12$, solve for a.

SOLUTION:

$\dfrac{3a}{3} = \dfrac{12}{3}$, dividing both sides by 3

$\dfrac{\cancel{3}a}{\cancel{3}} = 4$, cancelling; $\quad a = 4$

Check: $3(4) = 12; \quad 12 = 12$ $\checkmark$

As the equations become more complicated, several operations may have to be carried out in a single problem to solve the equation.

EXAMPLE:

$2x + 6 = x + 10$, solve for x.

SOLUTION:

$2x - x + 6 = x - x + 10$, subtracting x from both sides
$x + 6 = 10$, rewriting
$x + 6 - 6 = 10 - 6$, subtracting 6 from both sides
$x = 4$

Check: $2(4) + 6 = (4) + 10$
$8 + 6 = 14$; $14 = 14$ ✓

EXAMPLE:

$3x = x + 6$, solve for x.

SOLUTION:

$3x - x = x - x + 6$, subtracting x from both sides
$2x = 6$, rewriting
$\dfrac{\cancel{2}x}{\cancel{2}} = \dfrac{\cancel{6}\,3}{\cancel{2}}$, dividing both sides by 2 and cancelling
$x = 3$

Check: $3(3) = (3) + 6$; $9 = 9$ ✓

EXAMPLE:

$A = B/C$, solve for B.

SOLUTION:

$A \times C = \dfrac{B}{\cancel{C}} \times \cancel{C}$, multiplying both sides by C and cancelling
$A \times C = B$ or $B = AC$

Check: $A = \dfrac{A\cancel{C}}{\cancel{C}}$; $A = A$ ✓

As you can see, it doesn't matter whether numbers or letters are used in equations. As long as we treat each side of the equation the same, the equality will not be affected.

Let's include some parentheses in our equations and solve for the unknown quantities.

EXAMPLE:

$3(x + 6) = 42$, solve for x.

SOLUTION:

$$3x + 18 = 42$$
$$3x + 18 - 18 = 42 - 18; \quad 3x = 24$$
$$\frac{3x}{3} = \frac{24}{3}; \quad x = 8$$

EXAMPLE:

$\frac{y}{2} + 6 = -2(y + 9\frac{1}{2})$, solve for y.

SOLUTION:

$$\frac{y}{2} + 6 = -2y - 19$$
$$\frac{1y}{2} + 2y + 6 - 6 = -2y + 2y - 19 - 6 \qquad \text{Note: } \frac{y}{2} \text{ is the same as } \frac{1y}{2}$$
$$2\frac{1}{2}y = -25$$
$$\frac{5y}{2} = -25; \quad \frac{\frac{5y}{2}}{\frac{5}{2}} = \frac{-25}{\frac{5}{2}}$$
$$y = -\overset{5}{\cancel{25}} \times \frac{2}{\cancel{5}} = -10$$
$$y = -10$$

EXAMPLE:

a(b + c) = d, solve for a.

SOLUTION:

$$\frac{a\cancel{(b + c)}}{\cancel{(b + c)}} = \frac{d}{(b + c)} \text{ , dividing both sides by (b + c) and canceling}$$

$$a = \frac{d}{(b + c)}$$

EXAMPLE:

a(b + c) = d, solve for b.

SOLUTION:

ab + ac = d, removing parentheses

ab + ac − ac = d − ac, subtracting ac from both sides

ab = d − ac, rewriting

$$\frac{ab}{a} = \frac{(d - ac)}{a} \text{, dividing both sides by a}$$

$$b = \frac{(d - ac)}{a}$$

EXAMPLE:

12x − 5(x + 6) = 4x + 3, solve for x.

SOLUTION:

12x − 5x − 30 = 4x + 3

7x − 30 = 4x + 3

7x − 4x − 30 + 30 = 4x − 4x + 3 + 30

3x = 33

x = 11   (Check the answer)

TRANSPOSITION

In solving for x in the equation x − 4 = 8, we previously added a "+4" to both sides to get x by itself. Another way of achieving the same result would be to take the "-4", change its sign, and move it to the other side of the equation:

x − 4 = 8

x = 8 + 4 (This is the same as adding +4 to both sides)

x = 12

The process of moving a quantity from one side of an equation to the other by changing its sign of operation is called <u>transposition.</u>

In addition and subtraction, a quantity may be transposed from one side of an equation to another if its sign is changed.

A multiplier quantity may be removed from one side of an equation by making it a divisor in the other side. A divisor quantity may be removed from one side of an equation by making it a multiplier quantity in the other side.

EXAMPLE

$x + 6 = 14$
$x = 14 - 6$
$x = 8$

To get the x by itself, the +6 is transposed from the left to the right side of the equation by making it a "-6".

EXAMPLE

$a/4 = 7$
$a = 7 \times 4$
$a = 28$

To get "a" by itself, the divisor 4 on the left is changed to a multiplier 4 (4/1) on the right side of the equation.

EXAMPLE

$3b = 30$
$b = 30/3$
$b = 10$

To get "b" by itself, the multiplier 3 was changed to the divisor 3 on the right side of the equation.

For additional practice, solve the example problems in the previous few pages using transposition.

Transposition, then, is really just a shorter method for performing the basic addition, subtraction, multiplication and division operations on both sides of the equation.

You should use whichever method you like best for solving equations, however, transposition will shorten your solving times.

PRACTICE SET #17

Solve the following equations for the unknown symbol:

1. $x + 3 = 8$, $x = ?$

2. $2y = 10$, $y = ?$

3. $5c - 3 = 27$, $c = ?$

4. $3x + 5x = 64$

5. $\dfrac{5x}{2} = 25$

6. $57 = \dfrac{9b}{3}$

7. $18 = 5y - 2$

8. $\dfrac{n}{3} \times 6 = 2$

9. $3y = y + 8$

10. $2y + 2 = 14 - y$

11. $6(x - 3) + x = 24$

12. $\dfrac{1}{5} = \dfrac{1}{x}$

13. $a + a = 36$

14. $A = BC$, $C = ?$

15. $A = B + C$, $C = ?$

16. $V = \dfrac{A}{W}$, $W = ?$

17. $E = IR$, $R = ?$

18. $\dfrac{VR}{Vx} = \dfrac{R}{Rx}$, $R = ?$

19. $R = \dfrac{K \times L}{d^2}$, $L = ?$

20. $f = P \times \dfrac{N}{60}$, $N = ?$

Continued on next page

Continued from previous page

21. $R = \dfrac{2WH}{S + 0.1}$, $S = ?$

22. $R = \dfrac{2WH}{S + 0.1}$, $H = ?$

23. $R = \dfrac{2WH}{S + 0.1}$, $W = ?$

24. $P = 0.434 \times H$, $H = ?$

25. $H = P/AW$, $A = ?$

26. The perimeter of a rectangle is described by the formula $p = 2L + 2w$. If $L = 4'$ and $w = 3'$, find P.

27. The area of a triangle is found by the equation $A = \frac{1}{2}bh$. If $b = 4''$ and $h = 6''$, what is A?

28. If $C = \dfrac{(F - 32)5}{9}$ where C = temp. in °Centigrade and F = temp. in °Fahrenheit, find F if $C = 200°$.

29. If the pile driving equation is $R = \dfrac{2wh}{s + .1}$, find h when $R = 40{,}000$, $w = 8{,}000$ and $s = .7$.

30. If the equation for total resistance is $R_t = \dfrac{1}{\dfrac{1}{R_1} + \dfrac{1}{R_2} + \dfrac{1}{R_3}}$

    find R_t when $R_1 = 60$ ohms, $R_2 = 124$ ohms and $R_3 = 48$ ohms.

⬢⑨ *Powers and Roots*

If we wish to express the multiplication of 4 x 4, we can write it as 4^2, which is read "four squared" or "four to the second power". The superscript 2 above the 4 is called the exponent or <u>power</u> and indicates that the number 4 is to be used as a factor twice.

When we want to indicate that, say, the number 5 is to be used as a factor three times, we write "5^3", which is read "five cubed" or "five to the third power" and means 5 x 5 x 5.

If x = 2, then $x^3 = 8$ (2 x 2 x 2). If x = 2, then $x^4 = 16$ (2 x 2 x 2 x 2), and so on.

The <u>root</u> of a number is the opposite operation of raising a number to a power. Roots might be called equal factors of a number. 3 is called the <u>square root</u> of 9, since 3 x 3 = 9 and 3 is the number that multiplied by itself gives 9, or alternately, three squared is nine. Roots may be square roots, cube roots, fourth roots, and so on.

The sign indicating square root is $\sqrt{}$ called a radical or square root sign. It is placed over the number whose root is to be found. $\sqrt{16}$ means the "square root of 16", which is 4.

$$\sqrt{4} = 2 \qquad \sqrt{25} = 5 \qquad \sqrt{100} = 10$$

To show a root other than square root, an index or small number is placed over the radical sign.

$$\sqrt[3]{27} = \text{cube root of } 27 = 3 \quad (3 \times 3 \times 3 = 27)$$

$$\sqrt[4]{64} = 4$$

$$\sqrt[3]{1000} = 10$$

$$\sqrt[3]{16} = 2$$

To check to see if you have taken the correct root of a number, multiply it by itself the appropriate number of times. The product will equal the original number if the answer is correct.

The roots of the previous examples probably seem obvious. The roots of many numbers, however, cannot be determined at a glance. Several methods exist for finding the square and cube roots of any number.

Since time is usually an important factor in most exam situations, I highly recommend that you use math tables with squares, cubes, square roots and cube roots already solved for as many numbers as possible. The long-hand methods for finding roots may take several precious minutes, while a quick glance at the appropriate table will give the result in a matter of seconds. Tables for numbers .01 through 1000 are found in the back of this manual.

Just to familiarize yourself with using math tables, look up the following square roots:

$$\sqrt{.68} = \qquad \sqrt{25} = \qquad \sqrt{120} = \qquad \sqrt{756} =$$

The answers to the above square roots are .8246, 5, 10.9545 and 27.4955.

You may be thinking "the tables are fine if the numbers are between .01 and 1000. What if I need to find, say, the square root of 7400"?

Ok, good point. Let's examine several numbers below (taken from the tables in the back of this manual):

#1 Number	#2 Square	#3 Cube	#4 Square Root	#5 Cube Root
85	7225	614125	9.2195	4.3968
86	7396	636056	9.2736	4.4140
87	7569	658503	9.3274	4.4310
88	7744	681472	9.3808	4.4480
89	7921	704969	9.4340	4.4647

If we take the number 86 and square it (86 x 86), we obtain 7396. Then, conversely, the square root of 7396 is 86. Since 7400 is nearly the same as 7396, then the square root of 7400 is approximately

86. The table shows the relationship of these numbers both when squaring (see column #1 to column #2) as well as finding a square root (column #2 to column #1).

Let's summarize these steps:

> To approximate the square root of a number greater than 1000, using the tables in this manual,
> 1) Find the number in the column entitled SQUARE
> 2) Read back to the left one column — that number will be the approximate square root.

EXAMPLE:

Find the square root of 39,200.

SOLUTION:

1) page 87, 198^2
2) Thus, 198 is the approximate square root of 39,200.

EXAMPLE:

Find the square root of 673,000.

SOLUTION:

1) page 94, we see that

Number	Square
820	672,400
	673,000
821	674,041

The number 673,000 does not appear in the tables, but we can see that the square root of it will lie between 820 and 821. A close approximation would be 820.4.

Don't worry that the procedure shown above gives approximate square roots. The accuracy is more than adequate for competancy test purposes. Besides, most of the square roots, cube roots, etc. will involve the numbers smaller than 1000. This simple, yet effective way of handling the square root dilemma can save you valuable time on your examination. The same procedure can be applied to find cube roots of numbers larger than 1,000.

PRACTICE SET #18

Raise the following quantities to the indicated power:

1. $2^2 =$

2. $4^2 =$

3. $12^2 =$

4. $5^3 =$

5. 6 to the third power =

6. The fourth power of 3 =

7. $50^6 =$

8. $3^8 =$

9. $(1252)^2 =$

10. $(.02)^3 =$

11. The third power of .43 =

12. $(.0001)^4$

Find the following roots:

13. $\sqrt{16} =$

14. $\sqrt{81} =$

15. $\sqrt{100} =$

16. $\sqrt[3]{27} =$

17. $\sqrt[3]{64} =$

18. $\sqrt[4]{81} =$

19. $\sqrt[3]{1000} =$

20. $\sqrt{1.44} =$

21. $\sqrt{.0025} =$

22. $\sqrt{1} =$

23. $\sqrt{169} =$

24. $\sqrt{400} =$

25. $\sqrt{550} =$

26. $\sqrt{.14} =$

27. $\sqrt{.95} =$

28. $\sqrt{95} =$

29. $\sqrt{950} =$

30. $\sqrt{36,100} =$

31. $\sqrt{156,025} =$

32. $\sqrt[3]{27,000} =$

33. $\sqrt[3]{29,791} =$

34. $\sqrt{.0256} =$

35. $\sqrt[3]{321,419,125} =$

36. $\sqrt[3]{5,359,375} =$

37. $\sqrt{.8281} =$

38. $\sqrt{469,225} =$

39. $\sqrt{2304} =$

40. $\sqrt{9604} =$

◆⟨10⟩ *Board Feet*

Lumber's unit of measurement is called a board foot. This unit is 1 foot square by 1 inch thick.* For purposes of calculating board measure, lumber less than 1" thick is considered to be 1". The number of board feet in a piece of lumber is found by multiplying the length in feet by the width in inches by the thickness in inches and dividing the result by twelve.

$$\text{Board feet} = \frac{L \times w \times t}{12}, \quad \text{where } L = \text{length in feet}$$
$$w = \text{width in inches}$$
$$t = \text{thickness in inches}$$

EXAMPLE:

Find the number of board feet in a 2 x 6 that is 10 feet long.

SOLUTION:

$$\text{B.F.} = \frac{L \times w \times t}{12} = \frac{10 \times \cancel{6} \times \cancel{2}}{\cancel{12}_2} = 10 \text{ B.F.}$$

EXAMPLE:

How many B.F. are in (5) 6" x 8" that are 14' long?

SOLUTION:

For one 6 x 8, $\text{B.F.} = \dfrac{14 \times \cancel{6} \times \cancel{8}^4}{\cancel{12}_2} = 56 \text{ B.F}$

56 x 5 = 280 B.F.

To save you time, the board feet for a linear foot of lumber has been calculated and is shown in the back of this manual, page 81.

To calculate board feet using the table, simply find the board feet per linear foot value for the appropriate size and multiply by the total linear feet of lumber.

---

*These are "nominal dimensions." Due to trimming, planing, etc., the <u>actual</u> dimensions of a piece of lumber are less than nominal sizes.

EXAMPLE: Using the table in the back of this manual, how many board feet are in (10) 2" x 4"s that are 12 feet long?

SOLUTION:

Factor from table for 2 x 4 = .667 B.F./foot
Total linear feet = 10 x 12 = 120
120 x .667 = 80 B.F.

EXAMPLE: How many board feet are in (25) 2" x 10"s each being 12' long?

SOLUTION:

Factor from table for 2 x 10's = 1.667 B.F./foot
Total linear feet of 2 x 10 = 25 x 12' = 300'
300 x 1.667 = 500.1 Board Feet

PRACTICE SET #19

Find the number of board feet of lumber in the following problems:

1. 1 piece of 2" x 4" x 16' =

2. 5 pieces of 1" x 6" x 12' =

3. 6 pieces of 2" x 3" x 14' =

4. 28 pieces of 2" x 6" x 10' =

5. 30 pieces of 4" x 4" x 12' =

6. 12 pieces of 2" x 12" x 12' =

7. 10 pieces of 8" x 12" x 20' =

8. 25 pieces of 2" x 8" x 16' =

9. 18 pieces of 3" x 12" x 18' =

10. 524 pieces of 16" x 18" x 24' =

11. How many board feet are in 10 pieces that are $\frac{1}{2}$" x 3" x 12'?

12. How many board feet are in 250 linear feet of 1" x 3" Douglas fir?

13. 400 pieces of 1" x 6" each 12' long will cover a house floor area exactly. If 10% additional must be added for waste, how many total board feet must be ordered?

14. A built-up girder is to be made by fastening three 2" x 10" pieces together side by side to make a 6" x 10". If six of the girders are needed and each is 18' long, how many board feet of 2" x 10" are needed?

15. Determine the total cost of material below:

# pieces	t	w	length	Bd. ft.	Cost/Bd. ft.	Subcost
1	1"	12"	8'		$3.60	
4	2"	6"	14'		$3.40	
24	2"	4"	12'		$4.00	
8	2"	12"	18'		$5.20	_____
					Total Cost	

⬢⬢ ⑪ *Areas and Volumes*

There are various geometric figures, each with very specific characteristics and differing applications in the field. Points, lines, planes, and solids as well as combinations of these are very important and they each, naturally, have some mathematical equation associated with them. A surface or plane might be thought of as a sheet of paper having only two dimensions — length and width. A solid has three dimensions — length, width and depth.

An understanding of how to find the areas and volumes of various geometric figures may be essential if you are to calculate quantities of certain construction materials.

For example, paint, plaster, tile, roofing material and brick can be estimated by finding the total area these materials are to cover. Concrete, excavation and fill applications involve basic volume calculations of geometric solids.

The most important figures and formulas are explained in the following pages.

The symbols used in the formulas are:

A = area	s = side
SA = surface area	h = vertical height
V = volume	in^2 = sq. in. or square inches
p = perimeter	in^3 = cu. in. or cubic inches
r = radius	ft^2 = sq. ft. or square feet
d = diameter	sy = square yards
c = circumference	cy = cubic yards

The labels of your answers are very important. Always make the units match — convert all dimensions to feet or all to inches.

Area = inches x inches = sq. in.; feet x feet = sq. ft., etc.

Volume = inches x inches x inches = cubic inches = in^3; feet x feet x feet = cubic feet = ft^3, etc.

AREAS

Circle: a curved line on which every point is equally distant (or equidistant) from one point within called the center.

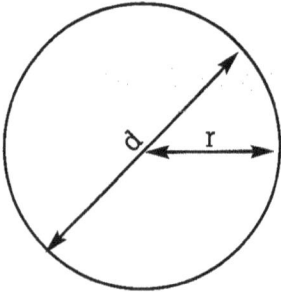

$A = \pi r^2$ (r = half the diameter)
Circumference (perimeter) $= 2\pi r = \pi d$

EXAMPLE:

For a circle having a radius of 4 inches, find the area and the circumference.

SOLUTION:

$A = \pi r^2 = 3.14(4)^2 = 3.14 \times 16 = 50.24$ in²
$C = 2\pi r = 2(3.14)(4) = 25.12$ inches

Triangle: a three-sided figure with sides that are straight lines.

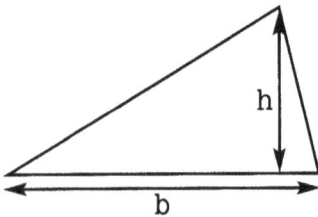

$A = \frac{1}{2}bh$

EXAMPLE:

Find the area of a triangle whose base is 8 inches and whose height is 5 inches.

SOLUTION:

$A = \frac{1}{2}bh = \frac{1}{2} \times 8 \times 5 = 40/2 = 20$ square inches

Before we continue with our review, a few notes on π. Pronounced "PIE", it is an infinite number roughly equal to 3.1415926535. More often than not, it is rounded to 3.14. This number represents a ratio of the circumference of a circle to its diameter. It has been historically referenced as early as 200 B.C. by the Ancient Egyptians and Babylonians. These early mathematicians estimated the value of π with relatively accurate approximations.

PYTHAGOREAN THEOREM

A <u>Right Triangle</u> is a triangle having one angle of 90°. The hypotenuse of a right triangle is the side opposite the right angle.

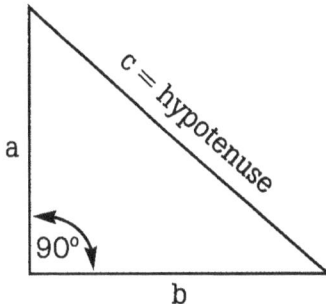

The square of the hypotenuse of a right triangle equals the sum of the squares of the other two sides.

$$c^2 = a^2 + b^2 \quad \text{or} \quad c = \sqrt{a^2 + b^2}$$
$$a^2 = c^2 - b^2 \quad \text{or} \quad a = \sqrt{c^2 - b^2}$$
$$b^2 = c^2 - a^2 \quad \text{or} \quad b = \sqrt{c^2 - a^2}$$

EXAMPLE:

The two short sides of a right triangle are 3 and 4 inches long. Find the length of the hypotenuse.

SOLUTION:

$c = \sqrt{a^2 + b^2}$
$\quad \sqrt{(3)^2 + (4)^2}$
$\quad \sqrt{9 + 16}$
$\quad \sqrt{25}$
$c = \quad 5$ inches

EXAMPLE:

If side b = 10" and side c = 15, find the length of side a.

SOLUTION:

$a = \sqrt{c^2 - b^2}$
$\quad \sqrt{(15)^2 - (10)^2}$
$\quad \sqrt{225 - 100} = \sqrt{125}$
$a = 11.18$ inches

AT A GLANCE

There are three basic triangle configurations each classified by the angles which appear at each of the three corners. All three angles of any triangle when added result in 180°. An acute triangle has no angle greater than 90°. An obtuse triangle has one angle greater than 90° and a right triangle has one angle exactly 90°. An isosceles triangle has two sides the same length while an equilateral triangle has all three sides or legs the same length.

Obtuse   Accute   Isosceles   Equilateral

Square: a geometric figure having four equal sides and four right angles.

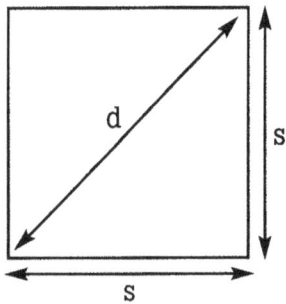

$A = s^2$
$p = 4s$
$\text{diagonal} = s\sqrt{2} = 1.414s$

EXAMPLE:

A square has sides that are 3 feet long. Find A, p and the diagonal.

SOLUTION:

$A = s^2 = (3)^2 = 3 \times 3 = 9$ sq. ft.
$p = 4s = 4 \times 3 = 12$ ft.
$\text{diagonal} = 1.414 \times 3' = 4.242$ ft.

Rectangle: a parallelogram all of whose angles are right angles.

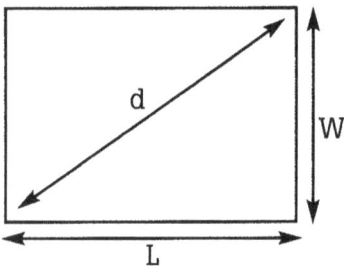

$A = LW$
$p = 2L + 2W$
$\text{diagonal} = \sqrt{L^2 + W^2}$

EXAMPLE:

A rectangle has sides with lengths of 4 and 6 inches. Find A, p and the diagonal.

SOLUTION:

$a = LW = 6 \times 4 = 24$ in.^{2}
$p = 2L + 2W = 2(6) + 2(4) = 12 + 8 = 20$ inches
$\text{diagonal} = \sqrt{L^2 + W^2} = \sqrt{(6)^2 + (4)^2} = \sqrt{36 + 16} = \sqrt{52} = 7.2$ inches

Parallelogram: a four-sided figure having opposite sides parallel and of equal length.

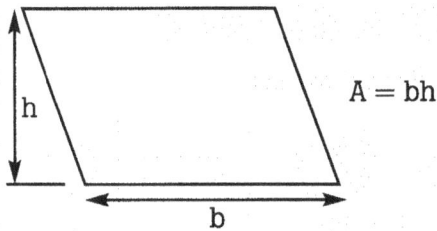

$A = bh$

EXAMPLE:

A parallelogram has a base of 8 inches and a vertical height of 7 inches. Find the area.

SOLUTION:

$A = bh = 8 \times 7 = 56$ in.^{2}

Trapezoid: a four-sided figure having two sides parallel.

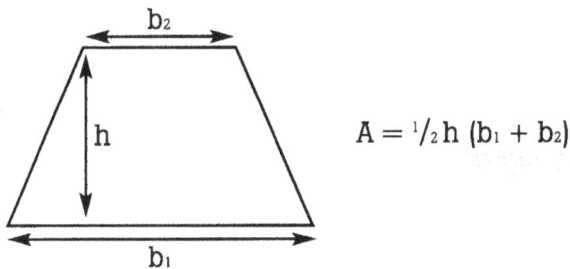

$A = \frac{1}{2} h (b_1 + b_2)$

EXAMPLE:

Find the area of a trapezoid having bases of 10" and 14" and a height of 7"

SOLUTION:

$A = \frac{1}{2} h (b_1 + b_2) = \frac{1}{2} \times 7(10 + 14)$
$= \frac{1}{2} \times 7(24) = \frac{1}{2} \times 168 = 84$ in^2

Two other regular polygons (closed plane figures bounded by straight lines) that are sometimes encountered are the hexagon and octagon. The formulas for the areas of these figures are given below.

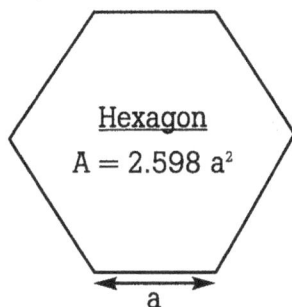

Hexagon
$A = 2.598 a^2$

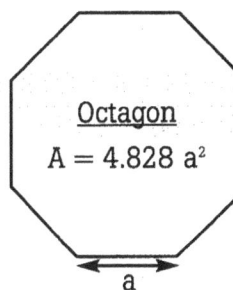

Octagon
$A = 4.828 a^2$

VOLUMES AND SURFACE AREAS OF SOLIDS

In the field, there are two main types of volumes that become part of our everyday construction experience. Rectangular solids and cylinders are the two most frequently used volumes. The Rectangular solid (or cube if all sides are the same) can be used to calculate concrete for footings, slabs or walks, air volume for rooms to help determine HVAC requirements, or for excavation when determining soil requirements for fill. The cylinder can be used for concrete calculations for pier footings or pedestals. Three other solid figures will be included in this chapter — the pyramid, cone and sphere — but they will seldom be seen in construction tests.

Cube: a rectangular solid with six equal square sides.

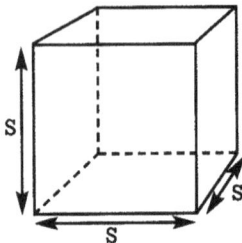

S.A. (surface area) = $6s^2$, where s = the dimension of one side
V = s^3

EXAMPLE:

Find the surface area (top, bottom and all four sides) and volume of a cube with a side of 5 inches.

SOLUTION:

S.A. = $6s^2$ = $6(5)^2$ = $6(25)$ = 150 sq. inches
Volume = s^3 = $(5)^3$ = 125 in.^{3}

Rectangular Solid: a geometric solid with six plane faces, straight edges and square corners.

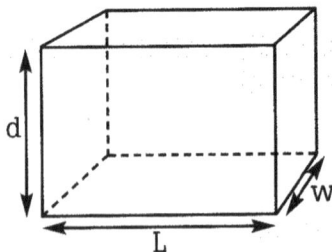

S.A. = $2(Lw + wd + Ld)$
V = Lwd

EXAMPLE:

Find the surface area and volume of a rectangular solid with dimensions of l = 9', w = 4', d = 3'.

SOLUTION:

S.A. = $2(Lw + wd + Ld)$ = $2(9 \cdot 4 + 4 \cdot 3 + 9 \cdot 3)$ = $2(36 + 12 + 27)$ = $2(75)$ = 150 sq. ft.
V = Lwd = (9)(4)(3) = 108 cubic feet

Cylinder:

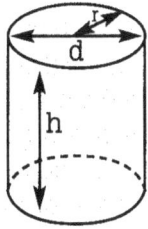

S.A. = $2\pi rh + 2\pi r^2$ (area of side plus area of top and bottom)

V = $\pi r^2 h$

EXAMPLE:

Find the surface area and volume of a cylinder 4' in diameter and 5' deep.

SOLUTION:

diameter = 2 x radius, so radius = $\frac{1}{2}$ diameter = $\frac{1}{2}$ x 4' = 2' = radius

S.A. = $2\pi rh + 2\pi r^2$ = [2(3.14)(2)(5)] + [2(3.14)(2)^2]

= 62.8 sq. ft. + 25.12 sq. ft = 87.92 sq. ft.

V = $\pi r^2 h$ = 3.14(2)^2(5) = 3.14(4)(5) = 62.8 cubic ft.

Right Circular Cone: a solid having a circle for a base and the axis of its height perpendicular to and passing through the center of its base.

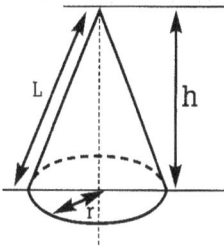

Lateral Surface Area = $\frac{1}{2}\pi dL$

Total Surface Area = $\frac{1}{2}\pi dL + \pi r^2$

Volume = $\frac{1}{3}Bh$, B = area of base = πr^2

AT A GLANCE

The lateral surface area is the "cone part" of the solid. Picture an empty ice cream cone — the part of the cone that holds the ice cream.

EXAMPLE:

Find the total surface area and volume of a right circular cone having a radius of 3", vertical height 4", and a slant height of 5" (slant height is the dimension "L" as shown in the drawing).

SOLUTION:

Total surface area = $\frac{1}{2}\pi dL + \pi r^2$ (d = diameter = 6")

= ($\frac{1}{2}$ x 3.14 x 6 x 5) + 3.14 x (3)^2 = 47.10 + 28.26 = 75.36 sq. in.

Volume = $\frac{1}{3}Bh = \frac{1}{3}\pi r^2$ x h = $\frac{1}{3}$ x 3.14 x (3)^2(4) = 37.68 cubic inches

The following geometrics are rarely seen in construction estimating and are seldom part of competancy exams. Becoming familiar with them is a good idea.

<u>Right Pyramids</u>: figures with polygons for bases (triangles, rectangles, squares, etc.) and points falling upon axes that are 90° angles from their base planes. The altitude of one of the triangles forming the sides or faces of the pyramid is called the <u>slant height</u>, the letter "h" in our example below. The <u>lateral surface area</u> of a pyramid is equal to the perimeter of its base multiplied by one half the slant height. The LSA is similar to the cone explanation on the previous page – a pyramid with no bottom. The <u>total surface area</u> of a pyramid is equal to the lateral surface area plus the area of the base.

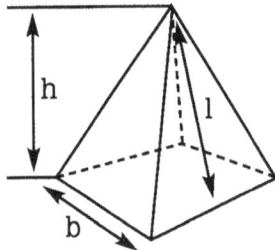

Area of face = lb/2
Lateral Surface Area = $\frac{1}{2}$ Sp (P = Perimeter of the base)
Total Surface Area = $\frac{1}{2}$ Sp + B (B = area of base)
Volume = $\frac{1}{3}$ Bh (h = vertical height)

EXAMPLE: Find the total surface area and volume of a pyramid whose base is square with sides of 4', vertical height of 6' and slant height of 6.32'.

SOLUTION:

Total surface area = $\frac{1}{2}$ lp + B = $\frac{1}{2}$ x 6.32 x (4 x 4) + (4)² = 50.56 + 16 = 66.56 sq. ft.
Volume = $\frac{1}{3}$ Bh = $\frac{1}{3}$ (4)² x 6 = 32 cu. ft.

<u>Sphere</u>: a perfectly round solid, such as a ball or globe.

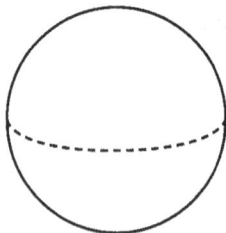

Surface Area = 4π r²
Volume = $\frac{4}{3}$ π r³

EXAMPLE: Find the surface area and volume of a sphere with a radius of 8 inches.
SOLUTION:

Surface area = 4π r² = 4 (3.14) (8)² = 803.84 in.²
Volume = $\frac{4}{3}$ π r³ = $\frac{4}{3}$ x 3.14 x 512 = 2138.2 in.³

PRACTICE SET #20

Find the areas of the following geometric figures:

1. A circle with a 6" radius.

2. A circle with a 10" radius.

3. A circle with a 12" diameter.

4. A circle with a 3' diameter.

5. A square with sides of 6".

6. A square with sides of 2'.

7. A rectangle with sides of 4" and 5".

8. A rectangle with sides of 3' and 6'.

9. A triangle with a base = 8" and height = 5".

10. A triangle with a base = 4" and height = 10".

11. A triangle with a base = 2' and height = 22'.

12. A trapezoid having bases of 8" and 10" and a height of 6".

13. A trapezoid having bases of 12" and 14" and a height of 8".

14. A parallelogram has bases of 10" and a height of 6".

15. A hexagon has sides of 5".

16. A hexagon has sides of 8'.

17. An octagon has sides of 5".

18. An octagon has sides of 2'.

Find the length of the third side of a right triangle knowing the following sides:

19. a = 3", b = 4", c = ?

20. a = 5', b = 8', c = ?

21. b = 12', c = 15', a = ?

22. a = 8", c = 12", b = ?

23. a = 15", b = 22", c = ?

24. b = 40", c = 50", a = ?

25. a = 3 $\frac{1}{2}$", b = 8 $\frac{1}{2}$", c = ?

26. a = 5.5', b = 10.75', c = ?

Continued on next page

Find the diagonal lengths for the following:

27. A square with sides of 5" long.

28. A rectangle with sides of 8' and 12'.

29. A square with sides of 12' long.

30. A rectangle with sides of 5" and 8".

Find the surface area and volume of the following geometric solids:

31. A cube with sides of 7".

32. A cube with sides of 10".

33. A rectangular solid with sides of 6", 8" and 12".

34. A rectangular solid with sides of 4', 6' and 10'.

35. A cylinder with a radius of 7" and height of 10".

36. A cylinder with a radius of 8' and height of 12'.

37. A cylinder with a radius of 16' and height of 12'.

38. A square-based pyramid with base side lengths of 6", a slant height of 10.44" and a vertical height of 10".

39. A rectangular-based pyramid with sides of 7" and 9", a slant height of 8.9" and a vertical height of 8".

40. A circular cone with a radius of 9", a slant height of 16.64" and a height of 14".

41. A sphere with a radius of 8".

Calculate the following:

42. Find the perimeter, area and diagonal distance of a square lot 200' on each side.

43. How many square feet of plywood will be needed to cover the tops of 150 square tables each 4' on a side?

44. Find the approximate number of $4\,^1/_2$" x $4\,^1/_2$" floor tiles for a floor 20' wide by 80' long.

45. Find the square feet of area of a garage gabled end (triangular) if the base is 24 feet and the height is 12 feet.

Continued on next page

46. What is the area of a trapezoid shaped lot that is 160' and 200' long (top and bottom) and 170' across?

47. Approximately how many hexagonal floor tile 2" on each side will be needed for a room that is 12' by 20'?

48. What is the area of a round tank bottom if its diameter is 21 feet?

49. What is the circumference of the above tank?

50. A roof has a rise of 4' and a run of 14'. What is the rafter length?

51. A stairs has a rise of 6' 3" and a run of 8' 6". What is the approximate length of the stair stringers used to support the steps?

52. A form for a concrete footing is 10' long, 22" wide and 10" deep. How many square feet of lumber are in this form (sides and ends only)?

53. How many gallons of paint would be needed to paint three rooms (walls only), each 12' by 14' by 8' high if the paint covers 500 sq. ft. per gallon?

54. Determine the number of rolls of paper for the above rooms if one roll has an effective coverage of 36 square feet.

55. How many cubic yards of concrete are in a circular slab 6" deep and with a diameter of 40'? (Divide cubic feet by 27 to find number of cubic yards.)

56. How many square feet of sheet metal will be required to line a cylindrical tank (top, bottom and sides) if it is 12' in diameter and 4' 8" high?

57. How many gallons will the above tank hold if there are 7.5 gallons per cubic foot?

58. How many square feet are in the surface of a spherical tank whose radius is 10'?

59. How many gallons would the above tank hold?

60. How many cubic yards of dirt is removed from an excavation 50' by 25' by 10' deep? (Neglect swell.)

61. A house that has dimensions of 24' by 40' long will have a flat roof with a 2' overhang all around. How many square feet of roof area is there?

62. If a roof with a 2' overhang all the way around and a rise of 8' and a run of 12' is to be built over the 40' long by 24' wide house above, how many square feet of roof area will there be?

63. A rectangular hangar is to house an airplane that is 120' long and 84' wide. A 20' allowance on all sides must be made to ensure adequate clearances and working space. What would the area of the hangar be?

64. How many cubic yards of asphaltic concrete paving would be needed for a runway 6500 ft. long, 100' wide and 8" deep? (Divide cubic feet by 27 to obtain cubic yards.)

Continued on next page

Continued from previous page

65. A rectangular concrete base 6' wide and 12' long supports two transformers whose bases are 42" by 28" each. What percent of the concrete area is actually covered by the transformers?

66. A ladder is to be used to reach the roof of a building that is 20' high. If 3' of the ladder must extend above the roof line, how long a ladder is needed? Assume that the bottom of the ladder will be 8' from the building.

67. A 40' long pine pole has an average diameter of 12". If pine weighs 42 lbs./cu. ft., how much does the pole weigh?

68. A concrete foundation wall is 80' long, 16" thick and 8' high. How many cubic yards of concrete are needed for the wall?

TABLE 1: Common Equivalent Units

1 mil = .001 inch

1 inch = 2.54 centimeters

12 inches = 1 foot

144 square inches = 1 square foot

1728 square inches = 1 cubic foot

3 feet = 1 yard

9 square feet = 1 square yard

1 cubic foot water = 62.4 pounds

27 cubic feet = 1 cubic yard

1760 yards = 1 mile

5280 feet = 1 mile

231 cubic inches = 1 gallon

7.48 gallons = 1 cubic foot

1 gallon (water) = 8.34 pounds

1 kilogram = 2.2 pounds

1 meter = 3.28 feet

1 ton = 2000 pounds

43,560 square foot = 1 acre

TABLE 2: Lineal Foot Table of Board Measure

Lumber Size		B.F. per Linear Foot
2" x 4"	=	0.667
2" x 6"	=	1
2" x 8"	=	1.333
2" x 10"	=	1.667
2" x 12"	=	2
2" x 14"	=	2.333
2" x 16"	=	2.667
3" x 6"	=	1.5
3" x 8"	=	2
3" x 10"	=	2.5
3" x 12"	=	3
3" x 14"	=	3.5
3" x 16"	=	4
4" x 4"	=	1.333
4" x 6"	=	2
4" x 8"	=	2.667
4" x 10"	=	3.333
4" x 12"	=	4
4" x 14"	=	4.667
4" x 16"	=	5.333
6" x 6"	=	3
6" x 8"	=	4
6" x 10"	=	5
6" x 12"	=	6
6" x 14"	=	7
6" x 16"	=	8
8" x 8"	=	5.333
8" x 10"	=	6.667
8" x 12"	=	8
8" x 14"	=	9.333
8" x 16"	=	10.667
10" x 10"	=	8.333
10" x 12"	=	10
10" x 14"	=	11.667
10" x 16"	=	13.333
10" x 18"	=	15
12" x 12"	=	12
12" x 14"	=	14
12" x 16"	=	16
12" x 18"	=	18
14" x 14"	=	16.333
14" x 16"	=	18.667
14" x 18"	=	21
16" x 16"	=	21.333
16" x 18"	=	24

TABLE 3: Fractions Converted to Decimals

Fraction	Decimal	Fraction	Decimal
$1/64$	0.015625	$33/64$	0.515625
$1/32$	0.03125	$17/32$	0.53125
$3/64$	0.046875	$35/64$	0.546875
$1/16$	0.0625	$9/16$	0.5625
$5/64$	0.078125	$37/64$	0.578125
$3/32$	0.09375	$19/32$	0.59375
$7/64$	0.109375	$39/64$	0.609375
$1/8$	0.125	$5/8$	0.625
$9/64$	0.140625	$41/64$	0.640625
$5/32$	0.15625	$21/32$	0.65625
$11/64$	0.171875	$43/64$	0.671875
$3/16$	0.1875	$11/16$	0.6875
$13/64$	0.203125	$45/64$	0.703125
$7/32$	0.21875	$23/32$	0.71875
$15/64$	0.234375	$47/64$	0.734375
$1/4$	0.25	$3/4$	0.75
$17/64$	0.265625	$49/64$	0.765625
$9/32$	0.28125	$25/32$	0.78125
$19/64$	0.296875	$51/64$	0.796875
$5/16$	0.3125	$13/16$	0.8125
$21/64$	0.328125	$53/64$	0.828125
$11/32$	0.34375	$27/32$	0.84375
$23/64$	0.359375	$55/64$	0.859375
$3/8$	0.375	$7/8$	0.875
$25/64$	0.390625	$57/64$	0.890625
$13/32$	0.40625	$29/32$	0.90625
$27/64$	0.421875	$59/64$	0.921875
$7/16$	0.4375	$15/16$	0.9375
$29/64$	0.453125	$61/64$	0.953125
$15/32$	0.46875	$31/32$	0.96875
$31/64$	0.484375	$63/64$	0.984375
$1/2$	0.5	1	1.0

<u>Caution:</u> do not confuse the values in this table with those in Table 5, page 84 (the figures in Table 5 are for fractional values of a <u>foot</u> expressed as a decimal equivalent).

Use the table on this page to simply find the decimal equivalent of a fraction.

EXAMPLE: Rewrite $7/32$" as a decimal.

SOLUTION: Table 3 value for $7/32$" = 0.21875

EXAMPLE: Express $7/32$" as the decimal equivalent of a foot.

SOLUTION: Table 5, p.84, under "0" inches and $7/32$",
$7/32$" = 0.0182'.

TABLE 4: Length of Common Rafters Per 12 Inches of Run

1 Pitch of Roof	2 Rise and Run or Cut	3 Length in Inches Common Rafter per 12" of Run	4 Multiply flat area of roof by
$1/12$	2 and 12	12.165	1.014
$1/8$	3 and 12	12.369	1.031
$1/6$	4 and 12	12.649	1.054
$5/24$	5 and 12	13.000	1.083
$1/4$	6 and 12	13.417	1.118
$7/24$	7 and 12	13.892	1.158
$1/3$	8 and 12	14.422	1.202
$3/8$	9 and 12	15.000	1.250
$5/12$	10 and 12	15.620	1.302
$11/24$	11 and 12	16.279	1.357
$1/2$	12 and 12	16.971	1.413
$13/24$	13 and 12	17.692	1.474
$7/12$	14 and 12	18.439	1.537
$5/8$	15 and 12	19.210	1.601
$2/3$	16 and 12	20.000	1.667
$17/24$	17 and 12	20.809	1.734
$3/4$	18 and 12	21.633	1.803
$19/24$	19 and 12	22.500	1.875
$5/6$	20 and 12	23.375	1.948
$7/8$	21 and 12	24.125	2.010
$11/24$	22 and 12	25.000	2.083
$11/12$	23 and 12	26.000	2.167
Full	24 and 12	26.875	2.240

+Use figures in this column to obtain area of roof for any pitch. To find the number of square feet of roof area where the pitch or rise and run is known, take the flat or horizontal area of the roof and multiply by the factor for that pitch. The result will be the roof area.

Any overhanging cornice may have to be added to the building area to obtain the total roof area.

EXAMPLE: What is the area of a roof 24' 0" by 38' 0" having an overhang of 2' 0" and a pitch of 1/8?

SOLUTION: Roof area 24' 0" + 2' 0" + 2' 0" = 28' 0" width
38' 0" + 2' 0" + 2' 0" = 42' 0" length
42 x 28 = 1176 sq. ft. horizontal area
To find area at 1/8 pitch, multiply flat area by appropriate factor (1.031)
1176 x 1.031 = 1212.5 sq. ft. of roof surface

TABLE 5: Decimals of a Foot (for each 32nd of an inch)

Inch	0	1	2	3	4	5	6	7	8	9	10	11
0	0	.0833	.1667	.2500	.3333	.4167	.5000	.5833	.6667	.7500	.8333	.9167
1/32	.0026	.0859	.1693	.2526	.3359	.4193	.5026	.5859	.6693	.7526	.8359	.9193
1/16	.0052	.0885	.1719	.2552	.3385	.4219	.5052	.5885	.6719	.7552	.8385	.9219
3/32	.0078	.0911	.1745	.2578	.3411	.4245	.5078	.5911	.6745	.7578	.8411	.9245
1/8	.0104	.0938	.1771	.2604	.3438	.4271	.5104	.5938	.6771	.7604	.8438	.9271
5/32	.0130	.0964	.1797	.2630	.3464	.4297	.5130	.5964	.6797	.7630	.8464	.9297
3/16	.0156	.0990	.1823	.2656	.3490	.4323	.5156	.5990	.6823	.7656	.8490	.9323
7/32	.0182	.1016	.1849	.2682	.3516	.4349	.5182	.6016	.6849	.7682	.8516	.9349
1/4	.0208	.1042	.1875	.2708	.3542	.4375	.5208	.6042	.6875	.7708	.8542	.9375
9/32	.0234	.1068	.1901	.2734	.3568	.4401	.5234	.6068	.6901	.7734	.8568	.9401
5/16	.0260	.1094	.1927	.2750	.3594	.4427	.5260	.6094	.6927	.7760	.8594	.9427
11/32	.0286	.1120	.1953	.2786	.3620	.4453	.5286	.6120	.6953	.7786	.8620	.9453
3/8	.0313	.1146	.1979	.2812	.3646	.4479	.5313	.6146	.6979	.7813	.8646	.9479
13/32	.0339	.1172	.2005	.2839	.3672	.4505	.5339	.6172	.7005	.7839	.8672	.9505
7/16	.0365	.1198	.2031	.2865	.3698	.4531	.5365	.6198	.7031	.7865	.8698	.9531
15/32	.0391	.1224	.2057	.2891	.3724	.4557	.5391	.6224	.7057	.7891	.8724	.9557
1/2	.0417	.1250	.2083	.2917	.3750	.4583	.5417	.6250	.7083	.7917	.8750	.9583
17/32	.0443	.1276	.2109	.2943	.3776	.4609	.5443	.6276	.7109	.7943	.8776	.9609
9/16	.0469	.1302	.2135	.2969	.3802	.4635	.5469	.6302	.7135	.7969	.8802	.9635
19/32	.0495	.1328	.2161	.2995	.3828	.4661	.5495	.6328	.7161	.7995	.8828	.9661
5/8	.0521	.1354	.2188	.3021	.3854	.4688	.5521	.6354	.7188	.8021	.8854	.9688
21/32	.0547	.1380	.2214	.3047	.3880	.4714	.5547	.6380	.7214	.8047	.8880	.9714
11/16	.0573	.1406	.2240	.3073	.3906	.4740	.5573	.6406	.7240	.8073	.8906	.9740
23/32	.0599	.1432	.2266	.3099	.3932	.4766	.5599	.6432	.7266	.8099	.8932	.9766
3/4	.0625	.1458	.2292	.3125	.3958	.4792	.5625	.6458	.7292	.8125	.8958	.9792
25/32	.0651	.1484	.2318	.3151	.3984	.4818	.5651	.6484	.7318	.8151	.8984	.9818
13/16	.0677	.1510	.2344	.3177	.4010	.4844	.5677	.6510	.7344	.8177	.9010	.9844
27/32	.0703	.1536	.2370	.3203	.4036	.4870	.5703	.6536	.7370	.8203	.9036	.9870
7/8	.0729	.1563	.2396	.3229	.4063	.4896	.5729	.6563	.7396	.8229	.9063	.9896
29/32	.0755	.1589	.2422	.3255	.4089	.4922	.5755	.6589	.7422	.8255	.9089	.9922
15/16	.0781	.1615	.2448	.3281	.4115	.4948	.5781	.6615	.7448	.8281	.9115	.9948
31/32	.0807	.1641	.2474	.3307	.4141	.4974	.5807	.6641	.7474	.8307	.9141	.9974

TABLE 6: Squares, Cubes, Square and Cube Roots

No.	Square	Cube	Square Root	Cube Root	No.	Square	Cube	Square Root	Cube Root
.01	.0001	.000001	0.1000	0.2154	.50	.2500	.125000	0.7071	0.7937
.02	.0004	.000008	0.1414	0.2714	.51	.2601	.132651	0.7141	0.7990
.03	.0009	.000027	0.1732	0.3107	.52	.2704	.140608	0.7211	0.8041
.04	.0016	.000064	0.2000	0.3420	.53	.2809	.148877	0.7280	0.8093
.05	.0025	.000125	0.2236	0.3684	.54	.2916	.157464	0.7348	0.8143
.06	.0036	.000216	0.2449	0.3915	.55	.3025	.166375	0.7416	0.8193
.07	.0049	.000343	0.2646	0.4121	.56	.3136	.175616	0.7483	0.8243
.08	.0064	.000512	0.2828	0.4309	.57	.3249	.185193	0.7550	0.8291
.09	.0081	.000729	0.3000	0.4481	.58	.3364	.195112	0.7616	0.8340
.10	.0100	.001000	0.3162	0.4642	.59	.3481	.205379	0.7681	0.8387
.11	.0121	.001331	0.3317	0.4791	.60	.3600	.216000	0.7746	0.8434
.12	.0144	.001728	0.3464	0.4932	.61	.3721	.226981	0.7810	0.8481
.13	.0169	.002197	0.3606	0.5066	.62	.3844	.238328	0.7874	0.8527
.14	.0196	.002744	0.3742	0.5192	.63	.3969	.250047	0.7937	0.8573
.15	.0225	.003375	0.3873	0.5313	.64	.4096	.262144	0.8000	0.8618
.16	.0256	.004096	0.4000	0.5429	.65	.4225	.274625	0.8062	0.8662
.17	.0289	.004913	0.4123	0.5540	.66	.4356	.287496	0.8124	0.8707
.18	.0324	.005832	0.4243	0.5646	.67	.4489	.300763	00.8185	0.8750
.19	.0361	.006859	0.4359	0.5749	.68	.4624	.314432	0.8246	0.8794
.20	.0400	.008000	0.4472	0.5848	.69	.4761	.328509	0.8307	0.8837
.21	.0441	.009261	0.4583	0.5944	.70	.4900	.343000	0.8367	0.8879
.22	.0484	.010648	0.4690	0.6037	.71	.5041	.357911	0.8426	0.8921
.23	.0529	.012167	0.4796	0.6127	.72	.5184	.373248	0.8485	0.8963
.24	.0576	.013824	0.4899	0.6214	.73	.5329	.389017	0.8544	0.9004
.25	.0625	.015625	0.5000	0.6300	.74	.5476	.405224	0.8602	0.9045
.26	.0676	.017576	0.5099	0.6383	.75	.5625	.421875	0.8660	0.9086
.27	.0729	.019683	0.5196	0.6463	.76	.5776	.438976	0.8718	0.9126
.28	.0784	.021952	0.5292	0.6542	.77	.5929	.456533	0.8775	0.9166
.29	.0841	.024389	0.5385	0.6619	.78	.6084	.474522	0.8832	0.9205
.30	.0900	.027000	0.5477	0.6694	.79	.6241	.493039	0.8888	0.9244
.31	.0961	.029791	0.5568	0.6768	.80	.6400	.512000	0.8944	0.9283
.32	.1024	.032768	0.5657	0.6840	.81	.6561	.531441	0.9000	0.9322
.33	.1089	.035937	0.5745	0.6910	.82	.6724	.551368	0.9055	0.9360
.34	.1156	.039304	0.5831	0.6980	.83	.6889	.571787	0.9110	0.9398
.35	.1225	.042875	0.5916	0.7047	.84	.7056	.592704	0.9165	0.9435
.36	.1296	.046656	0.6000	0.7114	.85	.7225	.614125	0.9220	0.9473
.37	.1369	.050653	0.6083	0.7179	.86	.7396	.636056	0.9274	0.9510
.38	.1444	.054872	0.6164	0.7243	.87	.7569	.658503	0.9327	0.9546
.39	.1521	.059319	0.6245	0.7306	.88	.7744	.681472	0.9381	0.9583
.40	.1600	.064000	0.6325	0.7368	.89	.7921	.704969	0.9434	0.9619
.41	.1681	.068921	0.6403	0.7429	.90	.8100	.729000	0.9487	0.9655
.42	.1764	.074088	0.6481	0.7489	.91	.8281	.753571	0.9539	0.9691
.43	.1849	.079507	0.6557	0.7548	.92	.8464	.778688	0.9592	0.9726
.44	.1936	.085184	0.6633	0.7606	.93	.8649	.804357	0.9644	0.9761
.45	.2025	.091125	0.6708	0.7663	.94	.8836	.830584	0.9695	0.9796
.46	.2116	.097336	0.6782	0.7719	.95	.9025	.857375	0.9747	0.9830
.47	.2209	.103823	0.6856	0.7775	.96	.9216	.884736	0.9798	0.9865
.48	.2304	.110592	0.6928	0.7830	.97	.9409	.912673	0.9849	0.9899
.49	.2401	.117649	0.7000	0.7884	.98	.9604	.941192	0.9899	0.9933
					.99	.9801	.970299	0.9950	0.9967

TABLE 6: Squares, Cubes, Square and Cube Roots (Continued)

No.	Square	Cube	Square Root	Cube Root	No.	Square	Cube	Square Root	Cube Root
1	1	1	1.0000	1.0000	50	2500	125000	7.0711	3.6840
2	4	8	1.4142	1.2599	51	2601	132651	7.1414	3.7084
3	9	27	1.7321	1.4422	52	2704	140608	7.2111	3.7325
4	16	64	2.0000	1.5874	53	2809	148877	7.2801	3.7563
5	25	125	2.2361	1.7100	54	2916	157464	7.3485	3.7798
6	36	216	2.4495	1.8171	55	3025	166375	7.4162	3.8030
7	49	343	2.6458	1.9129	56	3136	175616	7.4833	3.8259
8	64	512	2.8284	2.0000	57	3249	185193	7.5498	3.8485
9	81	729	3.0000	2.0801	58	3364	195112	7.6158	3.8709
10	100	1000	3.1623	2.1544	59	3481	205379	7.6811	3.8930
11	121	1331	3.3166	2.2240	60	3600	216000	7.7460	3.9149
12	144	1728	3.4641	2.2894	61	3721	226981	7.8102	3.9365
13	169	2197	3.6056	2.3513	62	3844	238328	7.8740	3.9579
14	196	2744	3.7417	2.4101	63	3969	250047	7.9373	3.9791
15	225	3375	3.8730	2.4662	64	4096	262144	8.0000	4.0000
16	256	4096	4.0000	2.5198	65	4225	274625	8.0623	4.0207
17	289	4913	4.1231	2.5713	66	4356	287496	8.1240	4.0412
18	324	5832	4.2426	2.6207	67	4489	300763	8.1854	4.0615
19	361	6859	4.3589	2.6684	68	4624	314432	8.2462	4.0817
20	400	8000	4.4721	2.7144	69	4761	328509	8.3066	4.1016
21	441	9261	4.5826	2.7589	70	4900	343000	8.3666	4.1213
22	484	10648	4.6904	2.8020	71	5041	357911	8.4261	4.1408
23	429	12167	4.7958	2.8439	72	5184	373248	8.4853	4.1602
24	576	13824	4.8990	2.8845	73	5329	389017	8.5440	4.1793
25	625	15625	5.0000	2.9240	74	5476	405224	8.6023	4.1983
26	676	17576	5.0990	2.9625	75	5625	421875	8.6603	4.2172
27	729	19683	5.1962	3.0000	76	5776	438976	8.7178	4.2358
28	784	21952	5.2915	3.0366	77	5929	456533	8.7750	4.2543
29	841	24389	5.3852	3.0723	78	6084	474552	8.8318	4.2727
30	900	27000	5.4772	3.1072	79	6241	493039	8.8882	4.2908
31	961	29791	5.5678	3.1414	80	6400	512000	8.9443	4.3089
32	1024	32768	5.6569	3.1748	81	6561	531441	9.0000	4.3267
33	1089	35937	5.7446	3.2075	82	6724	551368	9.0554	4.3445
34	1156	39304	5.8310	3.2396	83	6889	571787	9.1104	4.3621
35	1225	42875	5.9161	3.2711	84	7056	592704	9.1652	4.3795
36	1296	46656	6.0000	3.3019	85	7225	614125	9.2195	4.3968
37	1369	50653	6.0828	3.3322	86	7396	636056	9.2736	4.4140
38	1444	54872	6.1644	3.3620	87	7569	658503	9.3274	4.4310
39	1521	59319	6.2450	3.3912	88	7744	681472	9.3808	4.4480
40	1600	64000	6.3246	3.4200	89	7921	704969	9.4340	4.4647
41	1681	68921	6.4031	3.4482	90	8100	729000	9.4868	4.4814
42	1764	74088	6.4807	3.4760	91	8281	753571	9.5394	4.4979
43	1849	79507	6.5574	3.5034	92	8464	778688	9.5917	4.5144
44	1936	85184	6.6332	3.5303	93	8649	804357	9.6437	4.5307
45	2025	91125	6.7082	3.5569	94	8836	830584	9.6954	4.5468
46	2116	97336	6.7823	3.5830	95	9025	857375	9.7468	4.5629
47	2209	103823	6.8557	3.6088	96	9216	884736	9.7980	4.5789
48	2304	110592	6.9282	3.6342	97	9409	912673	9.8489	4.5947
49	2401	117649	7.0000	3.6593	98	9604	941192	9.8995	4.6104
					99	9801	970299	9.9499	4.6261

No.	Square	Cube	Square Root	Cube Root	No.	Square	Cube	Square Root	Cube Root
100	10000	1000000	10.0000	4.6416	150	22500	3375000	12.2474	5.3133
101	10201	1030301	10.0499	4.6570	151	22801	3442951	12.2882	5.3251
102	10404	1061208	10.0995	4.6723	152	23104	3511808	12.3288	5.3368
103	10609	1092727	10.1489	4.6875	153	23409	3581577	12.3693	5.3485
104	10816	1124864	10.1980	4.7027	154	23716	3652264	12.4097	5.3601
105	11025	1157625	10.2470	4.7177	155	24025	3723875	12.4499	5.3717
106	11236	1191016	10.2956	4.7326	156	24336	3796416	12.4900	5.3832
107	11449	1225043	10.3441	4.7475	157	24649	3869893	12.5300	5.3947
108	11664	1259712	10.3923	4.7326	158	24964	3944312	12.5698	5.4061
109	11881	1295029	10.4403	4.7769	159	25281	4019679	12.6095	5.4175
110	12100	1331000	10.4881	4.7914	160	25600	4096000	12.6491	5.4288
111	12321	1367631	10.5357	4.8059	161	25921	4173281	12.6886	5.4401
112	12544	1404928	10.5830	4.8203	162	26244	4251528	12.7279	5.4514
113	12769	1442897	10.6301	4.8346	163	26569	4330747	12.7671	5.4626
114	12996	1481544	10.6771	4.8488	164	26896	4410944	12.8062	5.4737
115	13225	1520875	10.7238	4.8629	165	27225	4492125	12.8452	5.4848
116	13456	1560896	10.7703	4.8770	166	27556	4574296	12.8841	5.4959
117	13689	1601613	10.8167	4.8910	167	27889	4657463	12.9228	5.5069
118	13924	1643032	10.8628	4.9049	168	28224	4741632	12.9615	5.5178
119	14161	1685159	10.9087	4.9187	169	28561	4826809	13.0000	5.5288
120	14400	1728000	10.9545	4.9324	170	28900	4913000	13.0384	5.5397
121	14641	1771561	11.0000	4.9461	171	29241	5000211	13.0767	5.5505
122	14884	1815848	11.0454	4.9597	172	29584	5088448	13.1149	5.5613
123	15129	1860867	11.0905	4.9732	173	29929	5177717	13.1529	5.5721
124	15376	1906624	11.1355	4.9866	174	30276	5268024	13.1909	5.5828
125	15625	1953125	11.1803	5.0000	175	30625	5359375	13.2288	5.5934
126	15876	2000376	11.2250	5.0133	176	30976	5451776	13.2665	5.6041
127	16129	2048383	11.2694	5.0265	177	31329	5545233	13.3041	5.6147
128	16384	2097152	11.3137	5.0397	178	31684	5639752	13.3417	5.6252
129	16641	2146689	11.3578	5.0528	179	32041	5735339	13.3791	5.6357
130	16900	2197000	11.4018	5.0658	180	32400	5832000	13.4164	5.6462
131	17161	2248091	11.4455	5.0788	181	32761	5929741	13.4536	5.6567
132	17424	2299968	11.4891	5.0916	182	33124	6028568	13.4907	5.6671
133	17689	2352637	11.5326	5.1045	183	33489	6128487	13.5277	5.6774
134	17956	2406104	11.5758	5.1172	184	33856	6229504	13.5647	5.6877
135	18225	2460375	11.6190	5.1299	185	34225	6331625	13.6015	5.6980
136	18496	2515456	11.6619	5.1426	186	34596	6434856	13.6382	5.7083
137	18769	2571353	11.7047	5.1551	187	34969	6539203	13.6748	5.7185
138	19044	2628072	11.7473	5.1676	188	35344	6644672	13.7113	5.7287
139	19321	2685619	11.7898	5.1801	189	35721	6751269	13.7477	5.7388
140	19600	2744000	11.8322	5.1925	190	36100	6859000	13.7840	5.7489
141	19881	2803221	11.8743	5.2048	191	36481	6967871	13.8203	5.7590
142	20164	2863288	11.9164	5.2171	192	36864	7077888	13.8564	5.7690
143	20449	2924207	11.9583	5.2293	193	37249	7189057	13.8924	5.7790
144	20736	2985984	12.0000	5.2415	194	37636	7301384	13.9284	5.7890
145	21025	3048625	12.0416	5.2536	195	38025	7414875	13.9642	5.7989
146	21316	3112136	12.0830	5.2656	196	38416	7529536	14.0000	5.8088
147	21609	3176523	12.1244	5.2776	197	38809	7645373	14.0357	5.8186
148	21904	3241792	12.1655	5.2896	198	39204	7762392	14.0712	5.8285
149	22201	3307949	12.2066	5.3015	199	39601	7880599	14.1067	5.8383

TABLE 6: Squares, Cubes, Square and Cube Roots (Continued)

No.	Square	Cube	Square Root	Cube Root	No.	Square	Cube	Square Root	Cube Root
200	40000	8000000	14.1421	5.8480	250	62500	15625000	15.8114	6.2996
201	40401	8120601	14.1774	5.8578	251	63001	15813251	15.8430	6.3080
202	40804	8242408	14.2127	5.8675	252	63504	16003008	15.8745	6.3164
203	41209	8365427	14.2478	5.8771	253	64009	16194277	15.9060	6.3247
204	41616	8489664	14.2829	5.8868	254	64516	16387064	15.9374	6.3330
205	42025	8615125	14.3178	5.8964	255	65025	16581375	15.9687	6.3413
206	42436	8741816	14.3527	5.9059	256	65536	16777216	16.0000	6.3496
207	42849	8869743	14.3875	5.9155	257	66049	16974593	16.0312	6.3579
208	43264	8998912	14.4222	5.9250	258	66564	17173512	16.0624	6.3661
209	43681	9129329	14.4568	5.9345	259	67081	17373979	16.0935	6.3743
210	44100	9261000	14.4914	5.9439	260	67600	17576000	16.1245	6.3825
211	44521	9393931	14.5258	5.9533	261	68121	17779581	16.1555	6.3907
212	44944	9528128	14.5602	5.9627	262	68644	17984728	16.1864	6.3988
213	45369	9663597	14.5945	5.9721	263	69169	18191447	16.2173	6.4070
214	45796	9800344	14.6287	5.9814	264	69696	18399744	16.2481	6.4151
215	46225	9938375	14.6629	5.9907	265	70225	18609625	16.2788	6.4232
216	46656	10077696	14.6969	6.0000	266	70756	18821096	16.3095	6.4312
217	47089	10218313	14.7309	6.0092	267	71289	19034163	16.3401	6.4393
218	47524	10360232	14.7648	6.0185	268	71824	19248832	16.3707	6.4473
219	47961	10503459	14.7986	6.0277	269	72361	19465109	16.4012	6.4553
220	48400	10648000	14.8324	6.0368	270	72900	19683000	16.4317	6.4633
221	48841	10793861	14.8661	6.0459	271	73441	19902511	16.4621	6.4713
222	49284	10941048	14.8997	6.0550	272	73984	20123648	16.4924	6.4792
223	49729	11089567	14.9332	6.0641	273	74529	20346417	16.5227	6.4872
224	50176	11239424	14.9666	6.0732	274	75076	20570824	16.5529	6.4951
225	50625	11390625	15.0000	6.0822	275	75625	20796875	16.5831	6.5030
226	51076	11543176	15.0333	6.0912	276	76176	21024576	16.6132	6.5108
227	51529	11697083	15.0665	6.1002	277	76729	21253933	16.6433	6.5187
228	51984	11852352	15.0997	6.1091	278	77284	21484952	16.6733	6.5265
229	52441	12008989	15.1327	6.1180	279	77841	21717639	16.7033	6.5343
230	52900	12167000	15.1658	6.1289	280	78400	21952000	16.7332	6.5421
231	53361	12326391	15.1987	6.1358	281	78961	22188041	16.7631	6.5499
232	53824	12487168	15.2315	6.1446	282	79524	22425768	16.7929	6.5577
233	54289	12649337	15.2643	6.1534	283	80089	22665187	16.8226	6.5654
234	54756	12812904	15.2971	6.1622	284	80656	22906304	16.8523	6.5731
235	55225	12977875	15.3297	6.1710	285	81225	23149125	16.8819	6.5808
236	55696	13144256	15.3623	6.1797	286	81796	23393656	16.9115	6.5885
237	56169	13312053	15.3948	6.1885	287	82369	23639903	16.9411	6.5962
238	56644	13481272	15.4272	6.1972	288	82944	23887872	16.9706	6.6039
239	57121	13651919	15.4596	6.2058	289	83521	24137569	17.0000	6.6115
240	57600	13824000	15.4919	6.2145	290	84100	24389000	17.0294	6.6191
241	58081	13997521	15.5242	6.2231	291	84681	24642171	17.0587	6.6267
242	58564	14172488	15.5563	6.2317	292	85264	24897088	17.0880	6.6343
243	59049	14348907	15.5885	6.2403	293	85849	25153757	17.1172	6.6419
244	59536	14526784	15.6205	6.2488	294	86436	25412184	17.1464	6.6494
245	60025	14706125	15.6525	6.2573	295	87025	25672375	17.1756	6.6569
246	60516	14886936	15.6844	6.2658	296	87616	25934336	17.2047	6.6644
247	61009	15069223	15.7162	6.2743	297	88209	26198073	17.2337	6.6719
248	61504	15252992	15.7480	6.2828	298	88804	26463592	17.2627	6.6794
249	62001	15438249	15.7797	6.2912	299	89401	26730899	17.2916	6.6869

TABLE 6: Squares, Cubes, Square and Cube Roots (Continued)

No.	Square	Cube	Square Root	Cube Root	No.	Square	Cube	Square Root	Cube Root
300	90000	27000000	17.3205	6.6943	350	122500	42875000	18.7083	7.0473
301	90601	27270901	17.3494	6.7018	351	123201	43243551	18.7350	7.0540
302	91204	27543608	17.3781	6.7092	352	123904	43614208	18.7617	7.0607
303	91809	27818127	17.4069	6.7166	353	124609	43986977	18.7883	7.0674
304	92416	28094464	17.4356	6.7240	354	125316	44361864	18.8149	7.0740
305	93025	28372625	17.4642	6.7313	355	126025	44738875	18.8414	7.0807
306	93636	28652616	17.4929	6.7387	356	126736	45118016	18.8680	7.0873
307	94249	28934443	14.5214	6.7460	357	127449	45499293	18.8944	7.0940
308	94864	29218112	17.5499	6.7533	358	128164	45882712	18.9209	7.1006
309	95481	29503629	17.5784	6.7606	359	128881	46268279	18.9473	7.1072
310	96100	29791000	17.6068	6.7679	360	129600	46656000	18.9737	7.1138
311	96721	30080231	17.6352	6.7752	361	130321	47045881	19.0000	7.1204
312	97344	30371328	17.6635	6.7824	362	131044	47437928	19.0263	7.1269
313	97969	30664297	17.6918	6.7897	363	131769	47832147	19.0526	7.1335
314	98596	30959144	17.7200	6.7969	364	132496	48228544	19.0788	7.1400
315	99225	31255875	17.7482	6.8041	365	133225	48627125	19.1050	7.1466
316	99856	31554496	17.7764	6.8113	366	133956	49027896	19.1311	7.1531
317	100489	31855013	17.8045	6.8185	367	134689	49430863	19.1572	7.1596
318	101124	32157432	17.8326	6.8256	368	135424	49836032	19.1833	7.1661
319	101761	32461759	17.8606	6.8328	369	136161	50243409	19.2094	7.1726
320	102400	32768000	17.8885	6.8399	370	136900	50653000	19.2354	7.1791
321	103041	33076161	17.9165	6.8470	371	137641	51064811	19.2614	7.1855
322	103684	33386248	17.9444	6.8541	372	138384	51478848	19.2873	7.1920
323	104329	33698267	17.9722	6.8612	373	139129	51895117	19.3132	7.1984
324	104976	34012224	18.0000	6.8683	374	139876	52313624	19.3391	7.2048
325	105625	34328125	18.0278	6.8753	375	140625	52734375	19.3649	7.2112
326	106276	34645976	18.0555	6.8824	376	141376	53157376	19.3907	7.2177
327	106929	34965783	18.0831	6.8894	377	142129	53582633	19.4165	7.2240
328	107584	35287552	18.1108	6.8964	378	142884	54010152	19.4422	7.2304
329	108241	35611289	18.1384	6.9034	379	143641	54439939	19.4679	7.2368
330	108900	35937000	18.1659	6.9104	380	144400	54872000	19.4936	7.2432
331	109561	36264691	18.1934	6.9174	381	145161	55306341	19.5192	7.2495
332	110224	36594368	18.2209	6.9244	382	145924	55742968	19.5448	7.2558
333	110889	36926037	18.2483	6.9313	383	146689	56181887	19.5704	7.2622
334	111556	37259704	18.2757	6.9382	384	147456	56623104	19.5959	7.2685
335	112225	37595375	18.3030	6.9451	385	148225	57066625	19.6214	7.2748
336	112896	37933056	18.3303	6.9521	386	148996	57512456	19.6469	7.2811
337	113569	38272753	18.3576	6.9589	387	149769	57960603	19.6723	7.2874
338	114244	38614472	18.3848	6.9658	388	150544	58411072	19.6977	7.2936
339	114921	38958219	18.4120	6.9727	389	151321	58863869	19.7231	7.2999
340	115600	39304000	18.4391	6.9795	390	152100	59319000	19.7484	7.3061
341	116281	39651821	18.4662	6.9864	391	152881	59776471	19.7737	7.3124
342	116964	40001688	18.4932	6.9932	392	153664	60236288	19.7990	7.3186
343	117649	40353607	18.5203	7.0000	393	154449	60698457	19.8248	7.3248
344	118336	40707584	18.5472	7.0068	394	155236	61162984	19.8494	7.3310
345	119025	41063625	18.5742	7.0136	395	156025	61629875	19.8746	7.3372
346	119716	41421736	18.6011	7.0203	396	156816	62099136	19.8997	7.3434
347	120409	41781923	18.6279	7.0271	397	157609	62570773	19.9249	7.3496
348	121104	42144192	18.6548	7.0338	398	158404	63044792	19.9499	7.3558
349	121801	42508549	18.6815	7.0406	399	159201	63521199	19.9750	7.3619

TABLE 6: Squares, Cubes, Square and Cube Roots *(Continued)*

No.	Square	Cube	Square Root	Cube Root	No.	Square	Cube	Square Root	Cube Root
400	160000	64000000	20.0000	7.3681	450	202500	91125000	21.2132	7.6631
401	160801	64481201	20.0250	7.3742	451	203401	91733851	21.2368	7.6688
402	161604	64964808	20.0499	7.3803	452	204304	92345408	21.2603	7.6744
403	162409	65450827	20.0749	7.3864	453	205209	92959677	21.2838	7.6801
404	163216	65939264	20.0998	7.3925	454	206116	93576664	21.3073	7.6857
405	164025	66430125	20.1246	7.3986	455	207025	94196375	21.3307	7.6914
406	164836	66923416	20.1494	7.4047	456	207936	94818816	21.3542	7.6970
407	165649	67419143	20.1742	7.4108	457	208849	95443993	21.3776	7.7026
408	166464	67917312	20.1990	7.4169	458	209764	96071912	21.4009	7.7082
409	167281	68417929	20.2237	7.4229	459	210681	96702579	21.4243	7.7138
410	168100	68921000	20.2485	7.4290	460	211600	97336000	21.4476	7.7194
411	168921	69426531	20.2731	7.4350	461	212521	97972181	21.4709	7.7250
412	169744	69934528	20.2978	7.4410	462	213444	98611128	21.4942	7.7306
413	170569	70444997	20.3224	7.4470	463	214369	99252847	21.5174	7.7362
414	171396	70957944	20.3470	7.4530	464	215296	99897344	21.5407	7.7418
415	172225	71473375	20.3715	7.4590	465	216225	100544625	21.5639	7.7473
416	173056	71991296	20.3961	7.4650	466	217156	101194696	21.5870	7.7529
417	173889	72511713	20.4206	7.4710	467	218089	101847563	21.6102	7.7584
418	174724	73034632	20.4450	7.4770	468	219024	102503232	21.6333	7.7639
419	175561	73560059	20.4695	7.4829	469	219961	103161709	21.6564	7.7695
420	176400	74088000	20.4939	7.4889	470	220900	103823000	21.6795	7.7750
421	177241	74618461	20.5183	7.4949	471	221841	104487111	21.7025	7.7805
422	178084	75151448	20.5426	7.5007	472	222784	105154048	21.7256	7.7860
423	178929	75686967	20.5670	7.5067	473	223729	105823817	21.7486	7.7915
424	179776	76225024	20.5913	7.5126	474	224676	106496424	21.7715	7.7970
425	180625	76765625	20.6155	7.5185	475	225625	107171875	21.7945	7.8025
426	181476	77308776	20.6398	7.5244	476	226576	107850176	21.8174	7.8079
427	182329	77854483	20.6640	7.5302	477	227529	108531333	21.8403	7.8134
428	183184	78402752	20.6882	7.5361	478	228484	109215352	21.8632	7.8188
429	184041	78953589	20.7123	7.5420	479	229441	109902239	21.8861	7.8243
430	184900	79507000	20.7364	7.5478	480	230400	110592000	21.9089	7.8297
431	185761	80062991	20.7605	7.5537	481	231361	111284641	21.9317	7.8352
432	186624	80621568	20.7846	7.5595	482	232324	111980168	21.9545	7.8406
433	187489	81182737	20.8087	7.5654	483	233289	112678587	21.9773	7.8460
434	188356	81746504	20.8327	7.5712	484	234256	113379904	22.0000	7.8514
435	189225	82312875	20.8567	7.5770	485	235225	114084125	22.0227	7.8568
436	190096	82881856	20.8806	5.5828	486	236196	114791256	22.0454	7.8622
437	190969	83453453	20.9045	7.5886	487	237169	115501303	22.0681	7.8676
438	191844	84027672	20.9284	7.5944	488	238144	116214272	22.0907	7.8730
439	192721	84604519	20.9523	7.6001	489	239121	116930169	22.1133	7.8784
440	193600	85184000	20.9762	7.6059	490	240100	117649000	22.1359	7.8837
441	194481	85766121	21.0000	7.6117	491	241081	118370771	22.1585	7.8891
442	195364	86350888	21.0238	7.6174	492	242064	119095488	22.1811	7.8944
443	196249	86938307	21.0476	7.6232	493	243049	119823157	22.2036	7.8998
444	197136	87528384	21.0713	7.6289	494	244036	120553784	22.2261	7.9051
445	198025	88121125	21.0950	7.6346	495	245025	121287375	22.2486	7.9105
446	198916	88716536	21.1187	7.6403	496	246016	122023936	22.2711	7.9158
447	199809	89314623	21.1424	7.6460	497	247009	122763473	22.2935	7.9211
448	200704	89915392	21.1660	7.6517	498	248004	123505992	22.3159	7.9264
449	201601	90518849	21.1896	7.6574	499	249001	124251499	22.3383	7.9317

TABLE 6: Squares, Cubes, Square and Cube Roots (Continued)

No.	Square	Cube	Square Root	Cube Root	No.	Square	Cube	Square Root	Cube Root
500	250000	125000000	22.3607	7.9370	550	302500	166375000	23.4521	8.1932
501	251001	125751501	22.3830	7.9423	551	303601	167284151	23.4734	8.1982
502	252004	126506008	22.4054	7.9476	552	304704	168196608	23.4947	8.2031
503	253009	127263527	22.4277	7.9528	553	305809	169112377	23.5160	8.2081
504	254016	128024064	22.4499	7.9581	554	306916	170031464	23.5372	8.2130
505	255025	128787625	22.4722	7.9634	555	308025	170953875	23.5584	8.2180
506	256036	129554216	22.4944	7.9686	556	309136	171879616	23.5797	8.2229
507	257049	130323843	22.5167	7.9739	557	310249	172808693	23.6008	8.2278
508	258064	131096512	22.5389	7.9791	558	311364	173741112	23.6220	8.2327
509	259081	131872229	22.5610	7.9843	559	312481	174676879	23.6432	8.2377
510	260100	132651000	22.5832	7.9896	560	313600	174616000	23.6643	8.2426
511	261121	133432831	22.6053	7.9948	561	314721	176558481	23.6854	8.2475
512	262144	134217728	22.6274	8.0000	562	315844	177504328	23.7065	8.2524
513	263169	135005697	22.6495	8.0052	563	316969	178453547	23.7276	8.2573
514	264196	135796744	22.6716	8.0104	564	318096	179406144	23.7487	8.2621
515	265225	136590875	22.6936	8.0156	565	319225	180362125	23.7697	8.2670
516	266256	137388096	22.7156	8.0208	566	320356	181321496	23.7908	8.2719
517	267289	138188413	22.7376	8.0260	567	321489	182284263	23.8118	8.2768
518	268324	138991832	22.7596	8.0311	568	322624	183250432	23.8328	8.2816
519	269361	139798359	22.7816	8.0363	569	323761	184220009	23.8537	8.2865
520	270400	140608000	22.8035	8.0415	570	324900	185193000	23.8747	8.2913
521	271441	141420761	22.8254	8.0466	571	326041	186169411	23.8956	8.2962
522	272484	142236648	22.8473	8.0517	572	327184	187149248	23.9165	8.3010
523	273529	143055667	22.8692	8.0569	573	328329	188132517	23.9374	8.3059
524	274576	143877824	22.8910	8.0620	574	329476	189119224	23.9583	8.3107
525	275625	144703125	22.9129	8.0671	575	330625	190109375	23.9792	8.3155
526	276676	145531576	22.9347	8.0723	576	331776	191102976	24.0000	8.3203
527	277729	146363183	22.9565	8.0774	577	332929	192100033	24.0208	8.3251
528	278784	147197952	22.9783	8.0825	578	334084	193100552	24.0416	8.3300
529	279841	148035889	23.0000	8.0876	579	335241	194104539	24.0624	8.3348
530	280900	148877000	23.0217	8.0927	580	336400	195112000	24.0832	8.3396
531	281961	149721291	23.0434	8.0978	581	337561	196122941	24.1039	8.3443
532	283024	150568768	23.0651	8.1028	582	338724	197137368	24.1247	8.3491
533	284089	151419437	23.0868	8.1079	583	339889	198155287	24.1454	8.3539
534	285156	152273304	23.1084	8.1130	584	341056	199176704	24.1661	8.3587
535	286225	153130375	23.1301	8.1180	585	342225	200201625	24.1868	8.3634
536	287296	153990656	23.1517	8.1231	586	343396	201230056	24.2074	8.3682
537	288369	154854153	23.1733	8.1281	587	344569	202262003	24.2281	8.3730
538	289444	155720872	23.1948	8.1332	588	345744	203297472	24.2487	8.3777
539	290521	156590819	23.2164	8.1382	589	346921	204336469	24.2693	8.3825
540	291600	157464000	23.2379	8.1433	590	348100	205379000	24.2899	8.3872
541	292681	158340421	23.2594	8.1483	591	349281	206425071	24.3105	8.3919
542	293764	159220088	23.2809	8.1533	592	350464	207474688	24.3311	8.3967
543	294849	160103007	23.3024	8.1583	593	351649	208527857	24.3516	8.4014
544	295936	160989184	23.3238	8.1633	594	352836	209584584	24.3721	8.4061
545	297025	161878625	23.3452	8.1683	595	354025	210644875	24.3926	8.4108
546	298116	162771336	23.3666	8.1733	596	355216	211708736	24.4131	8.4155
547	299209	163667323	23.3880	8.1783	597	356409	212776173	24.4336	8.4202
548	300304	164566592	23.4094	8.1833	598	357604	213847192	24.4540	8.4249
549	301401	165469149	23.4307	8.1882	599	358801	214921799	24.4745	8.4296

No.	Square	Cube	Square Root	Cube Root	No.	Square	Cube	Square Root	Cube Root
600	360000	216000000	24.4949	8.4343	650	422500	274625000	25.4951	8.6624
601	361201	217081801	24.5153	8.4390	651	423801	275894451	25.5147	8.6668
602	362404	218167208	24.5357	8.4437	652	425104	277167808	25.5345	8.6713
603	363609	219256227	24.5561	8.4484	653	426409	278445077	25.5539	8.6757
604	364816	220348864	24.5764	8.4530	654	427716	279726264	25.5734	8.6801
605	366025	221445125	24.5967	8.4577	655	429025	281011375	25.5930	8.6845
606	367236	222545016	24.6171	8.4623	656	430336	282300416	25.6125	8.6890
607	368449	223648543	24.6374	8.4670	657	431649	283593393	25.6320	8.6934
608	369664	224755712	24.6577	8.4716	658	432964	284890312	25.6515	8.6978
609	370881	225866529	24.6779	8.4763	659	434281	286191179	25.6710	8.7022
610	372100	226981000	24.6982	8.4809	660	435600	287496000	25.6905	8.7066
611	373321	228099131	24.7184	8.4856	661	436921	288804781	25.7099	8.7110
612	374544	229220928	24.7386	8.4902	662	438244	290117528	25.7294	8.7154
613	375769	230346397	24.7588	8.4948	663	439569	291434247	25.7488	8.7198
614	376996	231475544	24.7790	8.4994	664	440896	292754944	25.7682	8.7241
615	378225	232608375	24.7992	8.5040	665	442225	294079625	25.7876	8.7285
616	379456	233744896	24.8193	8.5086	666	443556	295408296	25.8070	8.7329
617	380689	234885113	24.8395	8.5132	667	444889	296740963	25.8263	8.7373
618	381924	236029032	24.8596	8.5178	668	446224	298077632	25.8457	8.7416
619	383161	237176659	24.8797	8.5224	669	447561	299418309	25.8650	8.7460
620	384400	238328000	24.8998	8.5270	670	448900	300763000	25.8844	8.7503
621	384641	239483061	24.9119	8.5316	671	450241	302111711	25.9037	8.7547
622	386884	240641848	24.9399	8.5362	672	451584	303464448	25.9230	8.7590
623	388129	241804367	24.9600	8.5408	673	452929	304821217	25.9422	8.7634
624	389376	242970624	24.9800	8.5453	674	454276	306182024	25.9615	8.7677
625	390625	244140625	25.0000	8.5499	675	455625	307546875	25.9808	8.7721
626	391876	245314376	25.0200	8.5544	676	456976	308915776	26.0000	8.7764
627	393129	246491883	25.0400	8.5590	677	458329	310288733	26.0102	8.7807
628	394384	247673152	25.0599	8.5635	678	459684	311665752	26.0384	8.7850
629	395641	248858189	25.0799	8.5681	679	461041	313046839	26.0576	8.7893
630	396900	250047000	25.0998	8.5726	680	462400	314432000	26.0768	8.7937
631	398161	251239591	25.1197	8.5772	681	463761	315821241	26.0960	8.7980
632	399424	252435968	25.1396	8.5817	682	465124	317214568	26.1151	8.8023
633	400689	253636137	25.1595	8.5862	683	466489	318611987	26.1343	8.8066
634	401956	254840104	25.1794	8.5907	684	467856	320013504	26.1534	8.8109
635	403225	256047875	25.1992	8.5952	685	469225	321419125	26.1725	8.8152
636	404496	257259456	25.2190	8.5997	686	470596	322828856	26.1916	8.8194
637	405769	258474853	25.2389	8.6043	687	471969	324242703	26.2107	8.8237
638	407044	259694072	25.2587	8.6088	688	473344	325660672	26.2298	8.8280
639	408321	260917119	25.2784	8.6132	689	474721	327082769	26.2488	8.8323
640	409600	262144000	25.2982	8.6177	690	476100	328509000	26.2679	8.8366
641	410881	263374721	25.3180	8.6222	691	477481	329939371	26.2869	8.8408
642	412164	264609288	25.3377	8.6267	692	478864	331373888	26.3059	8.8451
643	413449	265847707	25.3574	8.6312	693	480249	332812557	26.3249	8.8493
644	414736	267089984	25.3772	8.6357	694	481636	334255384	26.3439	8.8536
645	416025	268336125	25.3969	8.6401	695	483025	335702375	26.3629	8.8578
646	417316	269586136	25.4165	8.6446	696	484416	337153536	26.3818	8.8621
647	418609	270840023	25.4362	8.6490	697	485809	338608873	26.4008	8.8663
648	419904	272097792	25.4558	8.6535	698	487204	340068392	26.4197	8.8706
649	421201	273359449	25.4755	8.6579	699	488601	341532099	26.4386	8.8748

TABLE 6: Squares, Cubes, Square and Cube Roots (Continued)

No.	Square	Cube	Square Root	Cube Root	No.	Square	Cube	Square Root	Cube Root
700	490000	343000000	26.4575	8.8790	750	562500	421875000	27.3861	9.0856
701	491401	344472101	26.4764	8.8833	751	564001	423561751	27.4044	9.0896
702	492804	345948408	26.4953	8.8875	752	565504	425259008	27.4226	9.0937
703	494209	347428927	26.5141	8.8917	753	567009	426957777	27.4408	9.0977
704	495616	348913664	26.5330	8.8959	754	568516	428661064	27.4591	9.1017
705	497025	350402625	26.5518	8.9001	755	570025	430368875	27.4773	9.1057
706	498436	351895816	26.5707	8.9043	756	571536	432081216	27.4955	9.1098
707	499849	353393243	26.5895	8.9085	757	573049	433798093	27.5136	9.1138
708	501264	354894912	26.6083	8.9127	758	574564	435519512	27.5318	9.1178
709	502681	356400829	26.6271	8.9169	759	576081	437245479	27.5500	9.1218
710	504100	357911000	26.6458	8.9211	760	577600	438976000	27.5681	9.1258
711	505521	359425431	26.6646	8.9253	761	579121	440711081	27.5862	9.1298
712	506944	360944128	26.6833	8.9295	762	580644	442450728	27.6043	9.1338
713	508369	362467097	26.7021	8.9337	763	582169	444194947	27.6225	9.1378
714	509796	363994344	26.7208	8.9378	764	583696	445943744	27.6405	9.1418
715	511225	365525875	26.7395	8.9420	765	585225	447697125	27.6586	9.1458
716	512656	367061696	26.7582	8.9462	766	586756	449455096	27.6767	9.1498
717	514089	368601813	26.7769	8.9503	767	588289	451217663	27.6948	9.1537
718	515524	370146232	26.7955	8.9545	768	589824	452984832	27.7128	9.1577
719	516961	371694959	26.8142	8.9587	769	591361	454756609	27.7308	9.1617
720	518400	373248000	26.8328	8.9628	770	592900	456533000	27.7489	9.1657
721	519841	374805361	26.8514	8.9670	771	594441	458314011	27.7669	9.1696
722	521284	376367048	26.8701	8.9711	772	595984	460099648	27.7849	9.1736
723	522729	377933067	26.8887	8.9752	773	597529	461889917	27.8029	9.1775
724	524176	379503424	26.9072	8.9794	774	599076	463684824	27.8209	9.1815
725	525625	381078125	26.9258	8.9835	775	600625	465484375	27.8388	9.1855
726	527076	382657176	26.9444	8.9876	776	602176	467288576	27.8568	9.1894
727	528529	384240583	26.9629	8.9918	777	603729	469097433	27.8747	9.1933
728	529984	385828352	26.9815	8.9959	778	605284	470910952	27.8927	9.1973
729	531441	387420489	27.0000	9.0000	779	606841	472729139	27.9106	9.2012
730	532900	389017000	27.0185	9.0041	780	608400	474552000	27.9285	9.2052
731	534361	390617891	27.0370	9.0082	781	609961	476379541	27.9464	9.2091
732	535824	392223168	27.0555	9.0123	782	611524	478211768	27.9643	9.2130
733	537289	393832837	27.0740	9.0164	783	613089	480048687	27.9821	9.2170
734	538756	395446904	27.0924	9.0205	784	614656	481890304	28.0000	9.2209
735	540225	397065375	27.1109	9.0246	785	616225	483736625	28.0179	9.2248
736	541696	398688256	27.1293	9.0287	786	617796	485587656	28.0357	9.2287
737	543169	400315553	27.1477	9.0328	787	619369	487443403	28.0535	9.2326
738	544644	401947272	27.1662	9.0369	788	620944	489303872	28.0713	9.2365
739	546121	403583419	27.1846	9.0410	789	622521	491169069	28.0891	9.2404
740	547600	405224000	27.2029	9.0450	790	624100	493039000	28.1069	9.2443
741	549081	406869021	27.2213	9.0491	791	625681	494913671	28.1247	9.2482
742	550564	408518488	27.2397	9.0532	792	627264	493793088	28.1425	9.2521
743	552049	410172407	27.2580	9.0572	793	628849	498677257	28.1603	9.2560
744	553536	411830784	27.2764	9.0613	794	630436	500566184	28.1780	9.2599
745	555025	413493625	27.2947	9.0654	795	632025	502459875	28.1957	9.2638
746	556516	415160936	27.3130	9.0694	796	633616	504358336	28.2135	9.2677
747	558009	416832723	27.3313	9.0735	797	635209	506261573	28.2312	9.2716
748	559504	418508992	27.3496	9.0775	798	636804	508169592	28.2489	9.2754
749	561001	420189749	27.3679	9.0816	799	638401	510082399	28.2666	9.2793

No.	Square	Cube	Square Root	Cube Root	No.	Square	Cube	Square Root	Cube Root
800	640000	512000000	28.2843	9.2832	850	722500	614125000	29.1548	9.4727
801	641601	513922401	28.3019	9.2870	851	724201	616295051	29.1719	9.4764
802	643204	515849608	28.3196	9.2909	852	725904	618470208	29.1890	9.4801
803	644809	517781627	28.3373	9.2948	853	727609	620650477	29.2062	9.4838
804	646416	519718464	28.3549	9.2986	854	729316	622835864	29.2233	9.4875
805	648025	521660125	28.3725	9.3025	855	731025	625026375	29.2404	9.4912
806	649636	523606616	28.3901	9.3063	856	732736	627222016	29.2575	9.4949
807	651249	525557943	28.4077	9.3102	857	734449	629422793	29.2746	9.4986
808	652864	527514112	28.4253	9.3140	858	736164	631628712	29.2916	9.5023
809	654481	529475129	28.4429	9.3179	859	737881	633839779	29.3087	9.5060
810	656100	531441000	28.4605	9.3217	860	739600	636056000	29.3258	9.5097
811	657721	533411731	28.4781	9.3255	861	741621	638277381	29.3428	9.5134
812	659344	535387328	28.4956	9.3294	862	743044	640503928	29.3598	9.5171
813	660969	537367797	28.5132	9.3332	863	744769	642735647	29.3769	9.5207
814	662596	539353144	28.5307	9.3370	864	746496	644972544	29.3939	9.5244
815	664225	541343375	28.5482	9.3408	865	748225	647214625	29.4109	9.5281
816	665856	543338496	28.5657	9.3447	866	749956	649461896	29.4279	9.5317
817	667489	545338513	28.5832	9.3485	867	751689	651714363	29.4449	9.5354
818	669124	547343432	28.6007	9.3523	868	753424	653972032	29.4618	9.5391
819	670761	549353259	28.6182	9.3561	869	755161	656234909	29.4788	9.5427
820	672400	551368000	28.6356	9.3599	870	756900	658503000	29.4958	9.5464
821	674041	553387661	28.6531	9.3637	871	758641	660776311	29.5127	9.5501
822	675684	555412248	28.6705	9.3675	872	760384	663054848	29.5296	9.5537
823	677329	557441767	28.6880	9.3713	873	762129	665338617	29.5466	9.5574
824	678976	559476224	28.7054	9.3751	874	763876	667627624	29.5635	9.5610
825	680625	561515625	28.7228	9.3789	875	765625	669921875	29.5804	9.5647
826	682276	563559976	28.7402	9.3827	876	767376	672221376	29.5973	9.5683
827	683929	565609283	28.7576	9.3865	877	769129	674526133	29.6142	9.5719
828	685584	567663552	28.7750	9.3904	878	770884	676836152	29.6311	9.5756
829	687241	569722789	28.7924	9.3940	879	772641	679151439	29.6479	9.5792
830	688900	571787000	28.8097	9.3978	880	774400	681472000	29.6648	9.5828
831	690561	573856191	28.8271	9.4016	881	776161	683797841	29.6816	9.5865
832	692224	575930368	28.8444	9.4053	882	777924	686128968	29.6985	9.5901
833	693889	578009537	28.8617	9.4091	883	779689	688465387	29.7153	9.5937
834	695556	580093704	28.8791	9.4129	884	781456	690807104	29.7321	9.5973
835	697225	582182875	28.8964	9.4166	885	783225	693154125	29.7489	9.6010
836	698896	584277056	28.9137	9.4204	886	784996	695506456	29.7658	9.6046
837	700569	586376253	28.9310	9.4241	887	786769	697864103	29.7825	9.6082
838	702244	588480472	28.9482	9.4279	888	788544	700227072	29.7993	9.6118
839	703921	590589719	28.9655	9.4316	889	790321	702595369	29.8161	9.6154
840	705600	592704000	28.9828	9.4354	890	792100	704969000	29.8329	9.6190
841	707281	594823321	29.0000	9.4391	891	793881	707347971	29.8496	9.6226
842	708964	596947688	29.0172	9.4429	892	795664	709732288	29.8664	9.6262
843	710649	599077107	29.0345	9.4466	893	797449	712121957	29.8831	9.6298
844	712336	601211584	29.0517	9.4503	894	799236	714516984	29.8998	9.6334
845	714025	603351125	29.0689	9.4541	895	801025	716917375	29.9166	9.6370
846	715716	605495736	29.0861	9.4578	896	802816	719323136	29.9333	9.6406
847	717409	607645423	29.1033	9.4615	897	804609	721734273	29.9500	9.6442
848	719104	609800192	29.1204	9.4652	898	806404	724150792	29.9666	9.6477
849	720801	611960049	29.1376	9.4690	899	808201	726572699	29.9833	9.6513

TABLE 6: Squares, Cubes, Square and Cube Roots (Continued)

No.	Square	Cube	Square Root	Cube Root	No.	Square	Cube	Square Root	Cube Root
900	810000	729000000	30.0000	9.6549	950	902500	857375000	30.8221	9.8305
901	811801	731432701	30.0167	9.6585	951	904401	860085351	30.8383	9.8339
902	813604	733870808	30.0333	9.6620	952	906304	862801408	30.8545	9.8374
903	815409	736314327	30.0500	9.6656	953	908209	865523177	30.8707	9.8408
904	817216	738763264	30.0666	9.6692	954	910116	868250664	30.8869	9.8443
905	819025	741217625	30.0832	9.6727	955	912025	870983875	30.9031	9.8477
906	820836	743677416	30.0998	9.6763	956	913936	873722816	30.9192	9.8511
907	822649	746142643	30.1164	9.6799	957	915849	876467493	30.9354	9.8546
908	824464	748613312	30.1330	9.6834	958	917764	879217912	30.9516	9.8580
909	826281	751089429	30.1496	9.6870	959	919681	881974079	30.9677	9.8614
910	828100	753571000	30.1662	9.6905	960	921600	884736000	30.9839	9.8648
911	829921	756058031	30.1828	9.6941	961	923521	887503681	31.0000	9.8683
912	831744	758550528	30.1993	9.6976	962	925444	890277128	31.0161	9.8717
913	833569	761048497	30.2159	9.7012	963	927369	893056347	31.0322	9.8751
914	835396	763551944	30.2324	9.7047	964	929296	895841344	31.0483	9.8785
915	837225	766060875	30.2490	9.7082	965	931225	898632125	31.0644	9.8819
916	839056	768575296	30.2655	9.7118	966	933156	901428696	31.0805	9.8854
917	840889	771095213	30.2820	9.7153	967	935089	904231063	31.0966	9.8888
918	842724	773620632	30.2985	9.7188	968	937024	907039232	31.1127	9.8922
919	844561	776151559	30.3150	9.7224	969	938961	909853209	31.1288	9.8956
920	846400	778688000	30.3315	9.7259	970	940900	912673000	31.1448	9.8990
921	848241	781229961	30.3480	9.7294	971	942841	915498611	31.1609	9.9024
922	850084	783777448	30.3645	9.7329	972	944784	918330048	31.1769	9.9058
923	851929	786330467	30.3809	9.7364	973	946729	921167317	31.1929	9.9092
924	853776	788889024	30.3974	9.7400	974	948676	924010424	31.2090	9.9126
925	855625	791453125	30.4138	9.7435	975	950625	926859375	31.2250	9.9160
926	857476	794022776	30.4302	9.7470	976	952576	929714176	31.2410	9.9194
927	859329	796597983	30.4467	9.7505	977	954529	932574833	31.2570	9.9227
928	861184	799178752	30.4631	9.7540	978	956484	935441352	31.2730	9.9261
929	863041	801765089	30.4795	9.7575	979	958441	938313739	31.2890	9.9295
930	864900	804357000	30.4959	9.7610	980	960400	941192000	31.3050	9.9329
931	866761	806954491	30.5123	9.7645	981	962361	944076141	31.3209	9.9363
932	868624	809557568	30.5287	9.7680	982	964324	946966168	31.3369	9.9396
933	870489	812166237	30.5450	9.7715	983	966289	949862087	31.3528	9.9430
934	872356	814780504	30.5614	9.7750	984	968256	952763904	31.3688	9.9464
935	874225	817400375	30.5778	9.7785	985	970225	955671625	31.3847	9.9497
936	876096	820025856	30.5941	9.7819	986	972196	958585256	31.4006	9.9531
937	877969	822656953	30.6105	9.7854	987	974169	961504803	31.4166	9.9565
938	879844	825293672	30.6268	9.7889	988	976144	964430272	31.4325	9.9598
939	881721	827936019	30.6431	9.7924	989	978121	967361669	31.4484	9.9632
940	883600	830584000	30.6594	9.7959	990	980100	970299000	31.4643	9.9666
941	885481	833237621	30.6757	9.7993	991	982081	973242271	31.4802	9.9699
942	887364	835896888	30.6920	9.8028	992	984064	976191488	31.4960	9.9733
943	889249	838561807	30.7083	9.8063	993	986049	979146657	31.5119	9.9766
944	891136	841232384	30.7246	9.8097	994	988036	982107784	31.5278	9.9800
945	893025	843908625	30.7409	9.8132	995	990025	985074875	31.5436	9.9833
946	894916	846590536	30.7571	9.8167	996	992016	988047936	31.5595	9.9866
947	896809	849278123	30.7734	9.8201	997	994009	991026973	31.5753	9.9900
948	898704	851971392	30.7896	9.8236	998	996004	994011992	31.5911	9.9933
949	900601	854670349	30.8058	9.8270	999	998001	997002999	31.6070	9.9967
					1000	1000000	1000000000	31.6229	10.0000

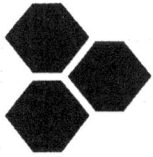

Answers

PRACTICE SET #1

1. $^3/_4$
2. $^5/_6$
3. $^3/_5$
4. $^1/_{14}$
5. $^6/_{11}$
6. $^1/_6$
7. $^1/_{10}$
8. $^1/_{250}$
9. $^1/_3$
10. $^1/_{11}$
11. 1
12. $^9/_{122}$ (already in lowest form)

PRACTICE SET #2

1. $4^1/_2$
2. 7
3. $2^1/_5$
4. $9^5/_9$
5. 16
6. $6^1/_2$
7. 2
8. 111
9. $1^1/_2$
10. $8^5/_{16}$
11. $^5/_2$
12. $^{28}/_5$
13. $^{60}/_7$
14. $^{38}/_3$
15. $^{207}/_{16}$
16. $^{94}/_7$
17. $^{75}/_4$
18. $^{207}/_8$
19. $^{509}/_{10}$
20. $^{311}/_3$

PRACTICE SET #3

1. $1^1/_2$
2. $1^1/_2$
3. $^3/_4$
4. $^4/_9$
5. $^7/_{20}$
6. $^7/_{12}$
7. $^{45}/_{52}$
8. 8
9. $10^{11}/_{12}$
10. $25^{13}/_{27}$
11. 5"
12. $27^1/_4$"
13. $1^1/_{64}$"
14. $^{185}/_{60} = 3^1/_{12}$ ohms
15. $^{174}/_{96} = 1^{13}/_{16}$
16. $43^7/_{24}$ lbs.

PRACTICE SET #4

1. 0
2. 1
3. $\frac{1}{4}$
4. $\frac{5}{16}$
5. $\frac{3}{8}$
6. $\frac{1}{2}$
7. $\frac{1}{16}$

8. $3\frac{2}{3}$
9. $1\frac{1}{64}$
10. $7\frac{8}{17}$
11. $1\frac{2}{3}$
12. $6\frac{13}{24}$
13. $397\frac{1}{8}$
14. $4922\frac{11}{12}$ lbs.

15. $\frac{3}{22}$"
16. $4\frac{1}{4}$"
17. $1\frac{23}{64}$"
18. $\frac{9}{32}$"
19. $8\frac{13}{16}$"
20. yes, $\frac{5}{6}$ hr. (50 minutes)

PRACTICE SET #5

1. $\frac{3}{25}$
2. $\frac{5}{14}$
3. $\frac{5}{9}$
4. $2\frac{5}{32}$
5. $\frac{2}{9}$
6. $27\frac{2}{3}$
7. 7650

8. $35\frac{3}{8}$
9. $\frac{7}{20}$
10. $95\frac{1}{5}$
11. $115\frac{5}{16}$" or 9' 7 $\frac{5}{16}$"
12. $11\frac{3}{8}$"
13. 13' 2 $\frac{7}{8}$"
14. 287 ft.

15. $4394\frac{1}{6}$ revolutions
16. 282,975 brick
17. 600 lbs.
18. $62\frac{1}{2}$ lbs.
19. $1362\frac{7}{12}$ lbs.

PRACTICE SET #6

1. $1\frac{1}{2}$
2. $\frac{1}{4}$
3. $5\frac{2}{5}$
4. 1
5. 8
6. $3\frac{5}{18}$
7. $7\frac{41}{42}$

8. $6\frac{102}{167}$
9. $39\frac{13}{28}$
10. $3\frac{7}{9}$
11. 816 sheets
12. 22 full pieces; $\frac{1}{3}$ ft.
13. 20 courses
14. 12 risers

15. $1\frac{41}{49}$ watts
16. 270 wedges (9 per strip)
17. $2\frac{3}{4}$"
18. $91\frac{4}{15}$ cubic feet
19. $8\frac{1}{3}$ pounds
20. 43 fittings

PRACTICE SET #7

1. .3
2. .6
3. .375
4. .4375
5. .078125
6. 2.25
7. 7.429 (rounded off)

8. 1.4
9. 701
10. 642.6
11. $^1/_4$
12. $^5/_8$
13. $^3/_{10}$
14. $^1/_{32}$

15. $^{13}/_{16}$
16. 2 $^1/_8$
17. 24 $^3/_8$
18. $^3/_{32}$
19. $^{13}/_{32}$
20. $^1/_{64}$

PRACTICE SET #8

1. .7
2. 26.57
3. 665.1809
4. 1.23
5. .0059

6. 5.426
7. 42.761
8. 1550.3113
9. $2719.79
10. 8.625"

11. .457"
12. 5.723 amps
13. $91.00
14. 4.25 hours
15. 1.835 ft.

PRACTICE SET #9

1. 6.8
2. 62.976
3. .026
4. .487706
5. 4504.5
6. 63
7. 68.46

8. 68.46
9. .064625
10. .0014965
11. 14,890,333.3333
12. .0029567
13. 227 loads
14. .403 lbs. (rounded off)

15. 2 inches
16. 455.06 lbs.
17. $123.75
18. 11.5 days
19. $82.80/sq. ft.
20. 29 pieces (28.8 exact)

PRACTICE SET #10

1. $\frac{1}{3}$' or .33'
2. $\frac{3}{4}$' or .75'
3. $\frac{3}{8}$' or .375'
4. .7396' (use decimals)
5. 1.4427' (use decimals)
6. 21.3516' (use decimals)
7. .4974' (use decimals)
8. .0807' (use decimals)
9. 120"
10. 63"
11. 315"
12. 8"
13. 103 $\frac{1}{2}$"
14. 1446"
15. 1 $\frac{1}{5}$" or 1.2"

PRACTICE SET #11

1. 12' 10"
2. 21' 2"
3. 19' 11 $\frac{3}{8}$"
4. 6' 9 $\frac{1}{8}$"
5. 4' 4 $\frac{5}{8}$"
6. 37' 2 $\frac{7}{16}$"
7. 24' 10"
8. 1 $\frac{3}{4}$"
9. 8.92" or 8' 11"
10. 8' 7 $\frac{1}{2}$"
11. 109.0'
12. 5.8828' (rounded off to 5.88')

PRACTICE SET #12

1. 10 sq. ft.
2. 48.21 sq. ft.
3. 39.67 sq. ft.
4. 36.73 sq. ft.
5. 5.25 cu. ft.
6. 50.72 cu. ft.
7. 4 ft.
8. 4.72 (rounded off)
9. 22.22"
10. 582"
11. 24.0
12. 4.828 (rounded off)
13. 41 brick per course
14. 27 pieces (27.42 exactly)
15. 10.44 ft.
16. 14 risers
17. 25 + 2 = 27 joist (24 spaces)
18. 16 pieces
19. approx. 43 poles (42 spaces with one on each end)
20. 22 pieces

PRACTICE SET #13

1. 50%	11. 17%	21. $280
2. 75%	12. 4.5%	22. 26 $^2/_3$
3. 130%	13. 120 gal.	23. 800 lbs.
4. 3%	14. 1.25 tons	24. $2040
5. .2%	15. 50%	25. 1000 brick
6. 25%	16. 20	26. 2750 board feet
7. 10%	17. $19,200	27. 1.25%
8. 2%	18. 25%	28. $55,080
9. 62.5%	19. 50	29. $631.80
10. 180%	20. 75%	30. $440

PRACTICE SET #14

1. 1 : 3	6. 2 : 25	10. $3000, $1500 and $500, respectively
2. 5 : 8	7. $^1/_3$	11. 380 lbs.
3. 3 : 20	8. $^1/_2$ (remember, run is $^1/_2$ span)	12. 9 to 100
4. 1 : 20	9. $^1/_2$	13. 400, 200, 120 and 80, respectively
5. 1 : 200		14. 32 qts. (4 qts. = 1 gal.)

PRACTICE SET #15

1. 6
2. 80
3. 4
4. 9
5. 1
6. 32
7. 200
8. 8
9. 1
10. 7
11. 8 gallons
12. $24.00
13. 250 gallons
14. 26,065 brick
15. 13 ft.
16. 1612 ft.
17. 7 cubic yards
18. 30 ft.
19. 164 fittings
20. 30.2 ft.
21. 650 ft.
22. 71.02 miles

PRACTICE SET #16

1. 12
2. 1
3. 20a
4. 4b
5. 11
6. 35
7. 10
8. 5a + 17
9. 9a + 9b
10. 6a + 3b
11. -12
12. -45
13. -72
14. $3\,\frac{1}{5}$
15. -88
16. 34a
17. 9
18. -3
19. -9
20. 0
21. 10
22. 14
23. 3
24. $6\,\frac{3}{4}$
25. 36
26. $6\,\frac{6}{7}$
27. -576
28. 0

PRACTICE SET #17

1. $x = 5$
2. $y = 5$
3. $c = 6$
4. $x = 8$
5. $x = 10$
6. $b = 19$
7. $y = 4$
8. $n = 1$
9. $y = 4$
10. $y = 4$
11. $x = 6$
12. $x = 5$
13. $a = 18$
14. $C = A/B$
15. $C = A - B$
16. $W = A/V$
17. $R = E/I$
18. $R = \dfrac{(VR)(Rx)}{Vx}$
19. $L = \dfrac{Rd^2}{K}$
20. $N = 60\,f/p$
21. $S = \dfrac{2WH - .1R}{R}$ or $S = \dfrac{2WH - 0.1}{R}$

(continued on next page)

PRACTICE SET #17 (Continued)

22. $H = \dfrac{Rs + .1R}{2W}$ or

$H = \dfrac{R(s + 0.1)}{2W}$

23. $W = \dfrac{Rs + .1R}{2H}$ or

$W = \dfrac{R(s + 0.1)}{2H}$

24. $H = \dfrac{P}{0.434}$

25. $A = \dfrac{P}{HW}$

26. $p = 14'$

27. $A = 12 \text{ in.}^2$

28. $F = \dfrac{9C}{5} + 32 = 392°$

29. $h = 2$

30. $\dfrac{357120}{16272} = 21.95 \text{ ohms}$

PRACTICE SET #18

1. 4
2. 16
3. 144
4. 125
5. 216
6. 81
7. 15,625,000,000
8. 6561
9. 1,567,504
10. .000008
11. .079507
12. .0000000000000001
13. 4
14. 9
15. 10
16. 3
17. 4
18. 3
19. 10
20. 1.2
21. .05
22. 1
23. 13
24. 20
25. 23.4521
26. .3742
27. .9747
28. 9.7468
29. 30.8221
30. 190
31. 395
32. 30
33. 31
34. .16
35. 685
36. 175
37. .91
38. 685
39. 48
40. 98

PRACTICE SET #19

1. 10.67 b.f.
2. 30 b.f.
3. 42 b.f.
4. 280 b.f.
5. 480 b.f.
6. 288 b.f.
7. 1600 b.f.
8. 533 $\frac{1}{3}$ b.f.
9. 972 b.f.
10. 301,824 b.f.
11. 30 b.f.
12. 62.5 b.f.
13. 2640 b.f.
14. 540 b.f.
15. $2484.84

PRACTICE SET #20 (Value of π used as 3.14)

1. 113.04 in.^{2}
2. 314 in.^{2}
3. 113.04 in.^{2}
4. 7.065 ft.^{2}
5. 36 sq. in.
6. 4 ft.^{2}
7. 20 sq. in.
8. 18 sq. ft.
9. 20 sq. in.
10. 20 sq. in.
11. 22 sq. in.
12. 54 sq. in.
13. 104 sq. in.
14. 60 in.^{2}
15. 64.95 in.^{2}
16. 166.27 ft.^{2}
17. 120.7 in^2
18. 19.31 ft.^{2}
19. 5"
20. 9.43'
21. 9"
22. 8.94"
23. 26.63 in.
24. 30"
25. approx. 9.19"
26. approx. 12.075"
27. 7.07"
28. 14.4 ft.
29. 16.97 ft.
30. 9.43"

(continued on next page)

31. Surface Area = 294 sq. in., V = 343 in.³

32. Surface Area = 600 sq. in., V = 1000 cu. in.

33. S.A. = 432 sq. in., V = 576 cu. in.

34. S.A. = 248 sq. ft., V = 240 cu. ft.

35. S.A. = 747.3 sq. in., V = 1538.6 cu. in.

36. S.A. = 1004.8 sq. ft., V = 2411.5 cu. ft.

37. S.A. = 2813.44 sq. ft., V = 9646.08 cu. ft.

38. S.A. = 161.28 sq. in., V = 120 cu. in.

39. S.A. = 205.4 sq. in., V = 168 cu. in.

40. S.A. = 724.6 sq. in., V = 1187 cu. in.

41. S.A. = 804 sq. in., V = 2144 cu. in.

42. p = 800', A = 40,000 sq. ft., d = 282.8'

43. 2400 sq. ft.

44. 11,378 tiles

45. 144 sq. ft.

46. 30,600 sq. ft.

47. 3326 tiles

48. 346.2 sq. ft.

49. 65.94'

50. 14.56'

51. Approx. 10 ½ ft

52. 19.7 sq. ft.

53. 2 ½ gal. (walls only)

54. 35 rolls needed

55. 23.26 cu. yd.

56. 402 sq. ft.

57. 3959 gallons

58. 1256 sq. ft.

59. 31,400 gallons

60. 463 cubic yards

61. 1232 sq. ft.

62. 1427 sq. ft.

63. 19,840 sq. ft.

64. 16,049 cu. yd.

65. 22.7%

66. 24 ½'

67. 1319 lbs.

68. 31.6 cu. yd.